APPLICATION
PROTOTYPING

APPLICATION PROTOTYPING

A REQUIREMENTS DEFINITION STRATEGY FOR THE 80s

BERNARD H. BOAR

Information Systems Consultant–
American Telephone & Telegraph Company

A Wiley-Interscience Publication

JOHN WILEY & SONS

New York   Chichester   Brisbane   Toronto   Singapore

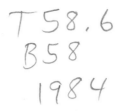

The opinions expressed in this book are those of the author
and do not necessarily represent the opinions of the American
Telephone and Telegraph Company.

Library of Congress Cataloging in Publication Data

Boar, Bernard H., 1947–
  Application prototyping.

  "A Wiley-Interscience publication."
  Includes index.
  1. Management information systems. 2. Business—
Data processing. 3. Decision-making. I. Title.
II. Title: Prototyping.
T58.6.B58 1984   658.4'012   83-16934
ISBN 0-471-89317-X

Printed in the United States of America

10 9 8 7

FOR DIANE, JESSICA, AND DEBBIE—WITH LOVE

CAVEAT

The management question, therefore, is not whether to build a pilot system and throw it away. You will do that. The only question is whether to plan in advance to build a throwaway, or to promise to deliver a throwaway to the customers. Seen this way, the answer is much clearer. Delivering that throwaway to the customer buys time, but it does so only at the cost of agony for the user, distraction for the builders while they do the redesign, and a bad reputation for the product that the best redesign will find hard to live down.

Hence plan to throw one away, you will anyhow.

FRED BROOKS, *THE MYTHICAL MAN MONTH*
ADDISON WESLEY, 1975, P. 116

PREFACE

It is disheartening and frustrating that despite all the grandiose proclamations of problem solution and noble efforts of users, analysts, and developers that requirements definition remains a highly imperfect process. In spite of repeated attempts to formalize and systematize the definition process, the creation of complete, correct, and consistent requirements remains an exceptional event. User complaints of "unworkable," "not what I wanted," or "incomplete" remain a recurring and common user reaction to delivered applications. The emperor unfortunately is naked. The current industry-endorsed prespecification methods for performing requirements definition often do not work.

There remain three outstanding and debilitating problems that all analysis techniques to date have failed to address adequately:

1 Users have extreme difficulty in prespecifying in total and final detail their requirements.
2 Graphical and narrative documentation techniques are inadequate to communicate the dynamics and ultimate acceptability of a proposed application.
3 Miscommunication is endemic between project members.

Currently used definition techniques such as structured analysis and specification languages have not been successful in solving these fundamental problems. An alternative technique is required that can directly solve these problems.

Requirements definition by the building of application models offers a viable and exciting solution to the definition problem. Prototyping is a method for extracting, presenting, and refining a user's needs by building a working model of the ultimate system—quickly and in context. By incrementally refining the model as mutual understanding of the problem and possible solutions evolve, prototyping can efficiently and effectively solve the definition problem.

The ability to deliver animated application models to the user changes everything about the definition process. No longer must the delivered system be a "surprise" or must project managers worry about misinterpretation. A working prototype provides physical anchors to both enable and enhance clear and unambiguous communication. Communication and debate can focus on active and dynamic system models as opposed to inanimate paper models.

The ability to deliver application models efficiently is a watershed event in data processing history. The resulting productivity improvement is in orders of magnitude. The real world problems of ultimate acceptability can be tested and verified in the "prototype laboratory" before an extensive commitment of development resources. The user can test the viability of a solution before committing to the application. Prototyping is the culminating event in the long and difficult effort to devise a "real world" workable definition strategy.

BERNARD H. BOAR

New Brunswick, New Jersey
November 1983

ACKNOWLEDGMENT

I would like to thank all the people who did the detail work of creating this text. Specifically, I would like to thank Leo Fath, a colleague and friend, who contributed immensely to the creation of the Prototyping Center at AT&T. Mr. Fath brought experience, insight, humor, and most of all common sense to our efforts.

B. H. B.

CONTENTS

ONE

INTRODUCTION TO APPLICATION PROTOTYPING

1.1 THE REQUIREMENTS QUESTION

The most pressing problem facing data processing management today is productivity. If you are serious about alleviating the productivity problem with application development, there is really only one question that requires your attention: What technique offers the highest probability of delivering a clear, correct, consistent, and validated requirement statement of a user's needs? Put all the buzzwords and politics aside for a while and attempt to directly answer that simple question.

There are many distractions that make it easy to avoid coming to grips with the problem. One's attention and concentration are easily and conveniently distracted by all the fog. Hardware vendors offer numerous appealing diversions: personal computers, distributed processing, dumb/smart/intelligent terminals, and local area networks. Software vendors provide query languages, relational data bases, nonprocedural report writers, interactive debuggers, code optimizers, program generators, and

1

dictionaries. Consultants provide an endless reservoir of advice and debate on structured life cycles, ergonomics, invisible data bases, software engineering, and the future of ADA.

The solutions offered by the vendors all help, but they really do not address the obvious. There is no efficiency, there is no effectiveness, there is no benefit, and there is no satisfaction if the business problem for which this arsenal of solutions has been assembled is not understood by all the interested parties in the first place.

If you regularly read the data processing literature, or even casually attend the seminar circuit, you are no doubt swimming (sinking?) with advice on the "golden" keys that unlock massive productivity improvement. Articles with titles like these are typical:

Program Generators Key to Programmer Productivity

Information Center Unlocks Door to Effectiveness

Personal Computer Key to User Satisfaction

Purchased Software Opens Door to Efficiency

Do you literally believe the claims? Do you truly believe that the fundamental problem with developing applications will be solved by simply buying another piece of hardware or software?

The problem cannot be the absence of good software. There are many fine tools available that ease and simplify the development task. The function delivered per line of code has increased by at least 100 to 1 in the last decade for users who have adopted improved software technology. The advent of fully integrated Fourth Generation Software Implementation Systems (such as IDMS by Cullinet) will continue to deliver increased function per line of source code in the future.

This evaluation of software productivity contradicts the conventional wisdom that software productivity is abysmal. Conventional wisdom, however, often over simplifies the situation. Let us assume that over the last 30 years machine performance has increased by an order of 10^6. At the same time let us assume that using lines of code as the unit of measure, programmer productivity has increased superficially by only 15 to 1. Even if this were the case, the function delivered per line of code (which is all that really matters) has increased tremendously.

Consider a WRITE statement written in COBOL 10 years ago. Today, using a sophisticated data base management system, not only does that statement accomplish the WRITE but the following functionality is also delivered:

1 Relational pointers to other records in the data base are updated.
2 Before and after log records are written to an automated recovery file.
3 Performance statistics are automatically captured.
4 Record locks that were set to prevent concurrent updating are released.

Though in both cases the same line of code was written, the function delivered in the latter case is substantially greater. Though dwarfed by the improvement in hardware performance, when viewed from a functionality perspective, those companies that have exploited evolving software technology have also made tremendous strides.

The improvement in hardware efficiency requires little elaboration. Users have almost an endless array of processors, terminals, and networking schemes to choose from. All of this is delivered with an ever continuing decline in unit cost. It is conceivable that we soon will be approaching almost zero cost logic.

The question is repeated: What technique should be used to solicit, document, present, and verify the user's needs? When this question can satisfactorily be answered, the end result will be the solution of the fundamental problem of application development: requirements definition.

1.2 THE PROTOTYPING ANSWER

This book is about application prototyping. It provides a direct answer to the question. For most commercial business applications, requirements definition will almost always be best accomplished by building a working prototype of the subject system. It is recommended that a definition strategy be employed wherein a working model of the application is initially built as a baseline and then is used as an anchor for successive refinements until a solution to the users needs has been demonstrated and concurred with by all development participants. This type of approach will consistently yield a higher probability of implementation success than rigorous definition techniques that attempt to develop a complete and final requirements statement using only descriptive and graphical documentation techniques.

The goal of this book is to explain application prototyping at both the conceptual and operational level. First, the conceptual business case to justify using a prototyping strategy is presented. Guidelines are then provided on how to perform prototyping in a typical business setting. Most of the existing literature on prototyping tells you to go do it. This book will also explain how effective prototyping is done.

The book is organized into seven chapters as follows.

Chapter 2—The Business Case

Explains why the currently endorsed rigorous definition techniques often fail and why prototyping will work in their place.

Chapter 3—The Prototype Life Cycle

Explains how prototyping fits within the development life cycle and provides a specific methodology for building models.

Chapter 4—The Prototyping Center

Explains the resources needed in terms of staffing, hardware, software, education, and facilities to provide a prototyping service.

Chapter 5—Project Management

Explains the impact of prototyping on the project management function.

Chapter 6—Implementing Application Prototyping

Provides a model plan to introduce prototyping within an organization.

Chapter 7—Issues and Concerns

Provides answers to some of the common questions posed by users, project managers, and developers about prototyping.

Chapter 8—Epilogue

A short summary of the major conclusions about prototyping and application development.

1.3   AN INTRODUCTION TO PROTOTYPING

Most currently recommended methods for defining business system requirements are designed to establish a final, complete, consistent, and correct set of requirements before the system is designed, constructed, seen, or experienced by the user. Common and recurring industry experience indicates that despite the use of such rigorous techniques, in many cases users still reject applications as neither correct nor complete upon implementation. Consequently, expensive, time-consuming, and divisive rework is required to harmonize the original specification with the definitive test of actual operational needs. In the worse case, rather than retrofit the delivered system, it is abandoned. Developers may build and test against specifications but users accept or reject against current and actual operational realities.

The currently endorsed rigorous definition approaches such as structured analysis or problem definition languages are not successful for defining many business applications. There exists a large set of analysis problems where the uncertainty of the requirements, communication problems, or the personalities of the participants makes attempts at complete prespecification both inappropriate and ultimately ineffective. When a rigorous approach is incorrectly applied in such a situation, the result is not only expensive in cost and time, but as importantly, it is expensive in the frustration and demoralization of the participants. As illustrated in Figure 1.1, it is not surprising that many users, developers, and project managers view beginning rigorous definition as a high risk and unrewarding task rather than a potentially positive undertaking.

An alternative approach to requirements definition is to capture an initial set of needs and to implement quickly those needs with the stated intent of iteratively expanding and refining them as mutual user/developer understanding of the system grows. Definition of the system occurs through gradual and evolutionary discovery as opposed to omniscient foresight. This approach assumes that project risk (make no mistake, software projects are high risk endeavors) is best controlled and minimized by using a technique that accepts gradual learning and incremental change as normal and desirable, and accommodates them efficiently by providing a technological "play-dough/tinker-toy" response. This kind of approach is called prototyping. It is also referred to as system modeling or heuristic development. It offers an attractive and workable alternative to

Figure 1.1 Rigorous definition. It is often viewed as a negative and unrewarding activity by project participants.

prespecification methods to deal better with the uncertainty, ambiguity, and fickleness of real world software projects.

Many prespecification techniques have their origins in dealing successfully with "toy" problems. Toy problems have the following characteristics:

They are small (most flow diagram examples fit neatly on a page).

An exact specification is provided.

A correct solution is readily determined that achieves instant consensus by all parties.

The participants, analyst, user, and developer are all high quality and are solely dedicated to the task.

Day-to-day software projects do not exhibit "toy" problem characteristics:

They are big (too big).

Specifications are imprecise and volatile.

The correctness of a solution is in the eye of each beholder.

The participants vary in skill level and are often multitasked.

It is not surprising that forward looking developers are questioning their methods and wondering if a better approach is available.

1.4   DEFINITIONS

As the terms "rigorous definition" (prespecification) and "application prototyping" are used extensively throughout the book, it is beneficial to clearly define and distinguish them:

Rigorous Definition.   Refers to a strategy for determining the business requirements of an application by prespecifying in detail all the needs prior to any contextual design, implementation, or operational experience. Techniques such as

- ☐   interviewing
- ☐   observation
- ☐   review of existing procedures and systems
- ☐   research of relevant business policies
- ☐   brainstorming

are all used to derive a logical model of the proposed system. Physical models are generally shunned as being inappropriate to the analysis phase. Modern analysis tools such as structured analysis and problem definition languages are commonly used to aid in analyzing, documenting, and presenting the requirements.

Application Prototyping.   Refers to a strategy for performing requirements determination wherein user needs are extracted, presented, and developed by building a working model of the ultimate system—quickly and in context. Though the same techniques as applied in a rigorous approach can be used to obtain an initial understanding of the problem, the approach differs significantly in that as soon as an initial understanding is made, an attempt is made to implement quickly that understanding. The first cut model now serves as a communication anchor between all

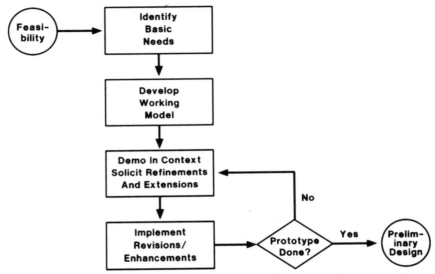

Figure 1.2   Requirements definition by prototyping. The four step approach provides the user with the opportunity to experience the solution before agreeing to it.

parties both to enable and enhance a meaningful dialogue. The model is then gradually expanded and refined as the project participants increase their comprehension of the problem and possible solutions.

Figure 1.2 depicts a simplified model of the prototyping approach. The basic steps and their goals are as follows:

Identify Basic Needs.   Determine the fundamental goals and objectives of the application, major business problems to be solved by the system, the data elements, record relationships, and functions to be performed.

Develop Working Model.   Quickly build a working model that delivers the key items identified in the first step. Supplement the users request with good system building practices. It is important to deliver the first model quickly to maintain user interest and confidence in the process.

Demonstrate in Context/Solicit Refinements and Extensions.   Present the model to all interested parties, from data entry clerk to division manager and aggressively solicit additional requirements. Attempt to mimic performance of the business service to discover shortcomings and desirable extensions. Walk through each component of the prototype explaining exactly what it does. Prompt the user by asking "what if" questions.

Prototype Done.　Continue to iterate between demonstrating and revising until the functionality provided is satisfactory to and understood by all parties. Test the prototype by attempting to perform the business service.

This distinction between prespecification and prototyping is important. They represent divergent philosophies on how to approach optimally the requirements definition problem. Those organizations that choose wisely will experience superior results.

1.5　DATA BASE NOTATION

A number of examples of application data bases are illustrated in the text. The following diagramming notation is used to illustrate record relationships:

1.　A rectangle with a record name within it (Figure 1.3, top left) is used to illustrate an instance of a record entity.

2.　Two record entities connected by an arrow (Figure 1.3, bottom left) is used to illustrate a "one-to-many" relationship between record entity types where the record at the tip of the arrow is the "many" partner in the relationship. Example: A "Company" has many "Divisions."

3.　A "many-to-many" relationship between record entities is illustrated by a record entity involved in multiple one-to-many relationships concurrently (Figure 1.3, top right). The record at the point of all the arrows is called a junction or intersection record. In some cases, there is meaningful data stored in the junction record. In other cases, it is only a structural necessity to permit modeling of a many-to-many relationship. Example: A "Division" may be spread across multiple "Locations" and a "Location" may be a residence for multiple "Divisions."

4.　A special case of the many-to-many relationship is called a cyclic or "bill of materials" structure (Figure 1.3, bottom right). In this case, the many-to-many relationship is between two records of the same type. For

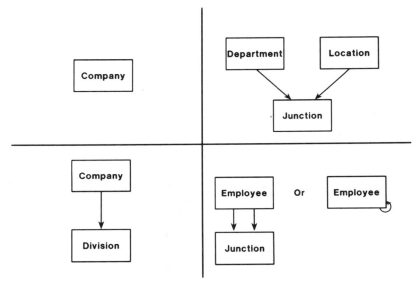

Figure 1.3 Data base diagramming notation. A record entity may: stand alone (top left), be engaged in a "one-to-many" relationship (bottom left), be engaged in a "many-to-many" relationship (top right), or be a special case of the "many-to-many" relationship called a cyclic structure (bottom right).

diagramming simplification, in some cases, cyclic structures will be illustrated by a looped arrow. Example: An "Employee" can be supervised by another employee and an "Employee" can supervise multiple other employees.

These four notations are the primitive constructs from which all data base diagrams in the book are derived. In practice a record entity can be concurrently involved in multiple relationships at the same time and in multiple roles (Figure 1.4).

1.6 PROTOTYPING AND APPLICATION SIZE

Most of the existing literature on rapid prototyping tends to "size" applications for which prototyping is appropriate as being small. A typical scenario would portray an analyst at a cathode-ray tube (CRT) meeting with a user. Two screens would be quickly generated. Following a quick

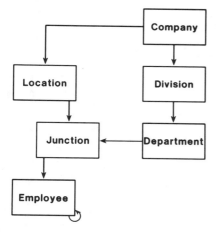

Figure 1.4 Composite data base diagram example. A record entity may be involved in multiple relationships of different types at the same time.

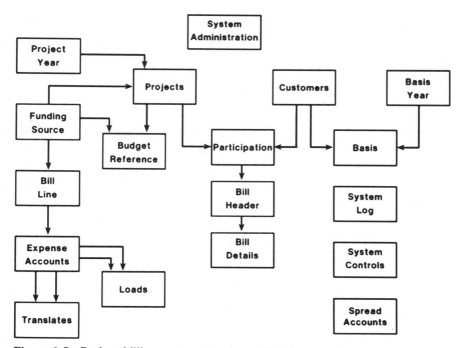

Figure 1.5 Project billing system data base. The larger and more complicated problems require prototyping much more than the small and simple ones.

iteration cycle, the user would be satisfied and the application would be completed for production implementation.

Given the state of software technology today, there is no reason why the techniques described in this book cannot be used to build rapid prototypes of medium and large size applications as well as the simple small ones. It is perfectly feasible to construct within a 4–6 week period, following the initial needs analysis, prototypes consisting of 25–35 screens and 40–60 on-line programs completely and perfectly documented.

Figure 1.5 (project billing system) illustrates an actual example in which these types of results were achieved. In 6 weeks, a team of two prototypers was able to build an initial model of the on-line maintenance and retrieval subsystems with the following deliverables:

- ☐ 33 "3270" type screens
- ☐ 67 on-line programs
- ☐ 200 logic modules that were "reused" by the 67 on-line programs
- ☐ 270 defined data elements
- ☐ 65 records
- ☐ a data base consisting of 23 record types and 29 inter-record relationships (not all illustrated)
- ☐ 53,000 lines of system documentation

This is obviously not a trivial problem nor can it be developed instantaneously in front of a user. Yet, it is illustrative of the type of problem prototyping is appropriate for. To achieve the maximum benefit from prototyping, the larger applications, not the smaller ones, should be considered as prototyping candidates.

TWO

THE
BUSINESS CASE

The purpose of this chapter is to develop the business arguments that support and justify the adoption of prototyping as a definition strategy. The best reason is simple: common sense. Daily life demonstrates the effectiveness of letting people resolve their needs by seeing and experiencing physical products. Users will respond to demonstrations by constructively critiquing what they see. Development environments need flexibility to evolve requirements. In fact, you would imagine that most data processing managers would go to sleep every night with only one wish: "May I be able to start building prototypes tomorrow." Nevertheless, a sound business case is the correct way to introduce a new idea and some people will have to be convinced.

This chapter will therefore provide the formal business case by developing the following topics:

- ☐ the critical importance of good requirements to system success
- ☐ a critical analysis of the rigorous definition strategy
- ☐ an analysis of the prototyping definition strategy
- ☐ the benefits of prototyping—both qualitative and quantitative
- ☐ the spectrum of definition environments

13

2.1   THE CRITICAL IMPORTANCE OF REQUIREMENTS

Most modern data processing installations perform their system development within the context of a structured development life cycle (SDLC) as
illustrated in Figure 2.1. An SDLC represents a graduated commitment to
application development with specific phases and phase deliverables to
permit project checkpoints, quality control, and restatement of management commitment and concurrence. The exact steps, their labels, and
deliverables vary between proprietary implementations. Figure 2.2 summarizes the generic goals and objectives of each phase of the representative SDLC (Figure 2.1).

All the structured life cycles, regardless of vendor, do agree on one
thing: the absolute importance of the definition step to project success. As
stated in the introduction, there will be no effectiveness, there will be no
efficiency, and there will be no benefit, if the problem is not clearly
understood in the first place.

In order to perform a requirements definition, it is necessary to know
the following types of things about an application:

Constraints.   Restrictions and limitations imposed on the applications by
the business environment, that is, predefined interfaces or policies of
noncompany entities such as the government.

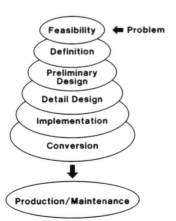

Figure 2.1   Structured development life cycle.
Most modern data processing development is
done within the context of a phased approach.

SDLC Phase	Goals And Objectives
Feasibility	Determine Technical, Operational, And Cost/Benefit Feasibility Of The Proposed Application
Definition	Determine The Requirements That The System Must Meet
Preliminary Design	Determine A Physical Solution To The Requirements
Detail Design	Provide An Exact Specification For The Construction Of Each System Component
Implementation	Construct, Test, And Verify The System
Conversion	Convert From The Current Mode Of Operation To The New System
Production	Operate The System On A Day-To-Day Basis
Maintenance	Revise The System To Maintain Operational Correctness

Figure 2.2   Goals and objectives of SDLC phases.

System Outputs.   A definition of each system output with its characteristics, that is, media, frequency, data element content, retention period, and so on.

System Inputs.   A definition of each system input with its characteristics, for example, data element content, source, volume, frequency, security considerations, and so on.

System Data Requirements.   A definition of the data within the system together with its group element relationships, logical record relationships, and inter-record relationships.

Data Elements. A definition of the characteristics and attributes of each data element, for example, format, name, synonyms, edit criteria, security, and so on.

Conversion (Cutover). How will the system go live? What is the source of start-up data and how will it populate the new system?

Function. A precise definition of what the system is to do. What are the logical transformations that the systems logic must perform, when are they performed, and on what data?

Controls/Auditing/Security. What measures are required to be put in the system to insure performance, data integrity, operational correctness, auditability, and security? How will errors be recycled and controlled within the system?

Performance/Reliability. What are the performance windows for the system? What level of downtime can be tolerated?

This list is by no means inclusive but it certainly makes the point: substantive requirements definition is a serious and exacting task. In addition, to be usable, the definition must have the following attributes; it must be:

Complete. All the requirements must be itemized.

Consistent. There should not be logical contradictions between requirements.

Nonredundant. There should not be wasted and confusing duplication.

Understandable. All parties should be able to interpret the requirements in a consistent manner. It must be clear and unambiguous as to the meaning of each requirement.

Testable. It must be possible to verify that the requirements are met.

Maintainable. The organization of the document must be such that it can be changed without disruption of the overall document.

Correct.   The requirements stated must be what is ultimately needed.

Necessary.   The requirements must be relevant. Vestigial requirements serve no purpose.

If it is incomplete, illogical, irrelevant, or misleading, regardless of the quality of the remaining steps, a disaster is inevitable. As currently performed requirements definition is precision work that to be of *any* utility must also be extremely high quality.

The almost universal emphasis placed on definition as the pivotal step to system success is readily understood. Numerous cost studies have demonstrated the exponential cost increase to correct errors or introduce additional function as you progress through the life cycle. Figure 2.3 provides an example of a typical "cost-to-correct" curve. Complimenting the cost-to-correct studies are a number of "source of system error" studies which show that 60–80% of all errors originate in definition (Figure 2.4). Consequently, development is faced with a double-edged sword. Not only is it ever more expensive to correct errors as we move through the life cycle, but also most of the errors originate in the early definition phase.

Though the cost to correct and the source of system error graphs are of more value as trend indicators than as precise measures, the industry

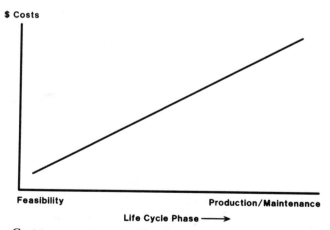

Figure 2.3   Cost-to-correct curve. The cost of correcting an error or revising system functionality grows exponentially as you progress through the life cycle.

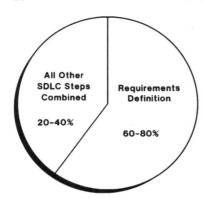

Figure 2.4 Source of system errors. The majority of all errors are traceable to poor requirements.

consensus is clear: You will pay even more dearly to revise your application as you move forward. If you are to control the system's cost and delivery schedule, it is imperative to enter preliminary design with a high degree of confidence that all parties understand and agree to the requirements.

As a result of this principle, there has been a justifiable interest in adopting definition techniques that promise to deliver high quality definitions. Developers have attempted to use "high-powered" prespecification techniques with complete methodologies to insure that a complete, consistent, and correct statement of requirements is generated.

Whether the rigorous approach has delivered in practice on its promise is controversial. What is not controversial is the importance of good requirements to project success. The problem confronting the time/cost driven developer who wishes to satisfy the user is clear: What strategy will permit me to obtain a definition statement with a low probability of change and a high probability of acceptance? Only a strategy that can deliver on these needs will be both effective and efficient.

2.2 RIGOROUS DEFINITION STRATEGY

The current industry endorsed method for performing requirements definition is a rigorous (prespecification) approach. In concept, a definition team attempts to thoroughly and completely prespecify the logical business needs of the application. Though the user is expected to review, critique, sign off, and live happily ever after with the proposed system, this must be done without the benefits of experience since the proposal normally only exists in graphical/narrative form. In fact, many structured analysis techniques specifically recommend that the specification should not be tainted by a suggested "physical" implementation.

Having created the logical vision of the future, the expectation is that the life cycle will proceed smoothly as illustrated in the representative SDLC (Figure 2.1). Recurring and common experience is quite to the contrary. Rather than the predicted orderly forward flow, many projects still experience extensive churning and rework to make the system operationally acceptable to the user. This unacceptability is first recognized at "system test" when the user gets to experience the system for the first time. In practice, prespecification does not consistently deliver an orderly life cycle.

Considering the promise and the hope offered by "rigor," one is tempted to ask the following question: Why are so many problems (user rejection, last minute rework, project cancellation) still encountered by projects that follow a rigorous discipline? Having adopted a prespecification strategy with religious intensity and complemented it with walk throughs, steering committees, project control systems, sign-offs, and multiple structured revolutions, why does the user still often reject the system at first touch as "unworkable," "not what I asked for," or "not what I wanted"? Why don't all the insurance policies payoff?

Proponents of rigorous definition will quickly argue that any failure is one of misexecution. The clean implementation and satisfied users were not obtained because of incomplete or insufficient rigor. The concept is correct, only the delivery system failed. Therefore, they will recommend even more rigor next time. In fact, this has led to the desirability of mechanized specification languages. They allegedly can mechanically analyze the specifications for completeness, correctness, and consistency. The defense is simple; we have provided the correct methodology, you must try harder to implement it correctly.

A more considered analysis of the problem indicates that throwing more and more rigor, more and more detail, and more and more specificity will also fail. Why?

2.2.1 Underlying Assumptions of Rigorous Definition

Figure 2.5 is an assumption/strategy pyramid. It is a convenient way to model decision making. At the foundation of the pyramid are assumed truths. Based on these "truths," interim logical conclusions are drawn; if

Figure 2.5 Assumption/strategy pyramid. Decision making can be viewed in terms of assumptions, conclusions, and strategies.

A therefore *B*. A strategy, the pinnacle of the pyramid, is then chosen to implement the conclusion; if *B*, I better do *C*. What is critical about this reasoning model is the dichotomy between the highly visible overt strategies on top and the often hidden assumptions at the bottom. The debates on the reasons for success or failure tend to focus only on the overt strategy. However, any strategy, no matter how well executed and noble in intent will fail if its underlying assumptions are incorrect. To understand why prespecification often fails, it is necessary to examine its supportive assumptions.

Figure 2.6 is the assumption/strategy pyramid that supports employing rigorous definition. When a project that uses prespecification falters, the postmortem analysis proceeds as follows:

> We have executed the strategy and it has not worked. The strategy is correct (by axiom). The failure is mine. I have misexecuted. Next time I will apply the strategy even more diligently.

The reasoning is perfect if we are content to limit ourselves to considering only the overt strategy as a point of postmortem review. What happens to the analysis if we expand the depth of our considerations to include questioning the underlying assumptions that support the strategy? What do we do if the assumptions are wrong?

The assumptions about a definition environment that support a rigorous approach are as follows:

1 All requirements can be prespecified.
2 Incomplete definition will be expensive to correct.
3 The project team is capable of unambiguous communication.

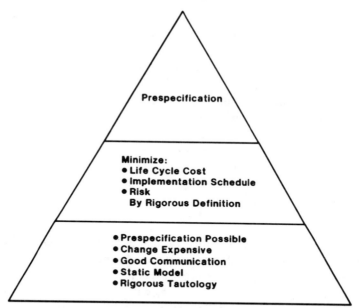

Figure 2.6 Assumption/strategy pyramid in support of prespecification. For a prespecification strategy to work, certain assumptions about the analysis environment must be true.

4 A static descriptive/graphical model of the proposed system is satisfactory.

5 A rigorous approach is inherently the correct approach for all life cycle phases.

If these assumptions are accepted as true, the remainder of the pyramid makes perfect sense. Life cycle cost, implementation schedule, and project risk will all be minimized by prespecifying the requirements in detail and totality. Before accepting this reasoning, however, we should carefully examine the assumptions to test their validity. Are we really guilty of misexecution or have we been attempting to follow automatically a strategy that is based on project-by-project sensitive assumptions?

Assumption: All Requirements Can Be Prespecified

This assumption must absolutely hold true if prespecification is to be viable. It asserts that all the needs can be completely itemized at the logical level without the aid of operational experience. The nuances and the subtleties can all be documented on paper. Can they? Though obviously in some situations they can, the human imperfections of the project participants often results in the fact that they cannot. Most people need to see examples and have practical experience before they are able to make judgments about the suitability of a proposed system and recommend revisions. Each individual involved in the project has a parochial view of what is needed. Joint consensus is most efficiently produced when a model is made available and discussion can proceed based on that model. People function superiorly in the role of discriminating consumer. People need to identify with something that exists in order to both confirm and refine their desires.

Even when a fine and valiant effort is made of prespecification, the initial contact with the solution changes individual's perception of what they want. They learn. Even the best prespecification will have to suffer through rework since the initial contact with the system will result in a learning experience by all participants and generate new insights into their needs. Needs evolve; they are not static. The users will reevaluate their original requests and want them changed for the better.

Change is equated with failure within the context of the prespecification mode. If it had been done right the first time (and it should have been) all this rework would not be necessary. The natural and inevitable act of evolving one's perception as one learns runs contradictory to the goals of prespecification.

Conclusion.   The validity of this assumption is highly tenuous. The people being called upon to provide specifications to the nth degree of detail are not professional specifiers nor do they find specifying interesting or exciting. It is extremely difficult to specify things of any complexity. Most users rarely ever specify anything; they pick and choose as an intelligent and discriminating consumer. Even when done well, prespecified systems will require rework and iteration because:

Individual and private visions of the system are not synchronized.

The experience of seeing and using the system stimulates the user to new desires.

Seeing and experiencing the system cancels previous requirements.

Assumption: Incomplete Definition Is Expensive

This assumption is based on the previously referenced "cost-to-correct" curve and "source of system errors" (Figures 2.3 and 2.4). It is apparently well grounded in industry experience and professional practitioner consensus. However, it has to be reevaluated in light of the availability of new software technology that makes the rapid building and changing of applications possible.

Why do children like play-dough, tinker-toys, and erector sets? Because they are flexible and easy to bend to the child's next whim. The child does not have to prespecify the idea; he merely needs a general idea of his vision and off he goes to work. If he does not like what he sees or a better idea pops into his head, a quick reordering of the tinker-toys or pounding of the play-dough is all that is required. The toys work in harmony with the child and permits him to use his imagination.

What if tinker-toy software existed? What if based on a limited statement of needs, objectives, or goals, software tinker-toys could be assembled and made operative. Based on seeing the "toy" system, components could be added, deleted, and replaced. Would not tinker-toy software permit cost efficient evolution of requirements and invalidate the "cost-to-correct" curve?

For now, let us make the following assertion: There exist software development systems that have evolved the concept of component engineering to the point where they provide a software equivalent to "tinker-toys." They provide sufficient flexibility, integration, leverage, and automation to invalidate this assumption. Incremental development of a systems requirements can take place within an operational model medium in a cost efficient manner.

Conclusion. Though it was true historically that due to the hardness of software that change was prohibitively labor intensive and expensive, the validity of this assumption must be reconsidered in light of modern software technology. Technology now exists which will permit software "tinker-toys" to be built and permit the efficient construction of software models.

Assumption: The Project Participants Are Capable
of Unambiguous Communication

As was the case with the initial assumption, this is a critical assumption if the rigorous approach is to work. In essence, it alleges that all the project team participants; project manager, analyst, user, developer, auditor, security analyst, data administrator, human factors specialist, and so on, are capable of clear and effective communication. Though each brings to the assignment a unique specialty, jargon, and perspective, the graphical/descriptive documentation tools will permit clear communication.

A user who does something routinely on a daily basis, automatically takes a lot for granted: "Of course they know that, doesn't everyone?" When the project participants read the definition report, do they translate it consistently? When a person describes the new system procedures at a walk through, do all the imaginations synchronize? Is there a meaningful dialogue or is there a hidden failure to communicate?

Each person goes ahead with their own interpretation. Is it really surprising that at implementation there are extensive discussions over what the specification said and didn't say? In practice, the specification said many things—what each person reads into it.

English is a very poor specification tool. Having recognized this, the prespecification techniques have offered Structured English, Tight English, decision tables, and tree diagrams as superior ways to communicate precisely. Though certainly an improvement over narrative English for minimizing ambiguity, they still lack the "rigor," "specificity," and "industry consensus" of a precision technical communication language.

Conclusion. It is difficult for multidisciplined individuals to bridge the communication gap. The failure to communicate effectively with each other has been recognized for a long time as one of the major failings of the development process. Though each person may overtly agree to a definition report, they often walk away with private and different understandings.

Assumption: A Static Descriptive/Graphical Model of the Application Is Adequate

The primary vehicle for communicating between people within a prespecification approach is the definition report (working and final). Though the exact format may vary between proprietary techniques, they all make use of a common set of primitives:

Narratives. Conventional English narrative explanations of goals, objectives, and other requirements.

Graphic Modeling. A flow diagramming technique to show the movement of data groups between external entities, processes, and files.

Logic Rules. Guidelines for presenting system transform logic in an unambiguous manner (decision tables).

Data Dictionary. A facility for defining all system entities, their attributes, and interentity relationships.

What is eminently consistent across all the techniques is that they are passive communicators. They are entirely static in nature and do not provide facilities for demonstrating the dynamics of the proposed application.

The assumption is that a passive model is adequate. Is it? Applications are becoming more and more interactive in nature. The distance between the user and the machine has shrunk. The requirements will result in "conversations" between the user and machine. Can a passive model hope to communicate the action of such a system?

The answer, of course, is no. Football or ice hockey can be explained to some degree by showing still pictures but to truly make an individual understand the excitement and movement of the game, they must be seen in motion. Similarly, to expect a user to make acceptance decisions based on still photos of an action packed system is unfair. There is a clear need for an "experiential" model.

It is not an accident that the rigorous models are passive. Their purpose is to provide a logical presentation of the system without the constraints of a physical solution. Though logical modeling is certainly beneficial and may be used in the early phases of a prototyping approach, it is incomplete. This is a fundamental difference between the two approaches. Prespecification places a great emphasis on passive logical models. Prototyping emphasizes active physical models.

Conclusion. Rigorous definition techniques are passive in nature. They consequently have difficulty portraying the "life" of an application. To be able to understand and critique a proposed application intelligently, it must be experienced; not just read and talked about.

Assumption: A Rigorous Approach Is Inherently Correct for All Life Cycle Phases

It has concerned many data processing practitioners that software development has been more akin to a craft than an engineering discipline.

There has been a worthy effort to develop ''software engineering'' principles that will escalate software development to a peer level with other engineering disciplines. Since most engineering techniques rely on strict methodology and rigorous discipline, so should all phases of software development. It is by axiom that ''rigor'' is correct for performing definition.

Is it? Requirements are often a fuzzy thing in the user's mind; this is especially true as one progresses from operational support systems (systems that are the business) to decision support systems (systems that are about the business). To insist on an exact and unforgiving statement of need from a person who has at best an embryonic idea is unworkable. They can be cornered to make a decision because the life cycle demands they do, but, if it were wrong, they will inevitably insist on it being revised. Flexibility, not rigidity, is needed to deal with the incomplete need.

It is absolutely necessary to be rigorous in construction once you in fact know the system you wish built. Rigor for design, coding, testing, change control, and so on, is certainly correct. But first we need an iterative ''good guess'' approach to get a solid statement of need. Requirement definition is a research project, not a manufacturing assembly line.

Conclusion. A rigorous approach cannot be assumed correct. It must compete with other approaches for legitimacy. Heuristic approaches are commonly used in other disciplines to test the validity of ideas. Prespecification should earn validity, not be granted it automatically.

2.2.2 Rigorous Definition Conclusion

In many cases, one or more of the underlying assumptions that legitimize prespecification are in fact not true. Consequently, with a weak or broken foundation, it is not surprising that the overt strategy fails and the project falters. Figure 2.7 demonstrates the consequences of applying prespecification to a definition problem where the assumptions are not true. Rather than the promised orderly forward flow exhibited in Figure 2.1, incessant looping and thrashing occur at a high cost in both dollars and people. The postmortem that automatically will recommend more rigor next time will yield but another expensive disappointment. The problem is not the overt strategy, the problem is the foundation assumptions.

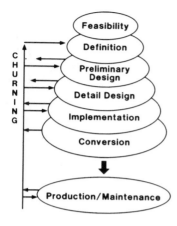

Figure 2.7 SDLC when supportive assumptions for prespecification fail. The absence of supporting assumptions results in expensive rework.

This analysis indicates that the majority of definition problems in conventional business settings to which a prespecification approach is routinely and automatically applied do not meet the prerequisite conditions. The assumptions that will make prespecification a valid approach are not true. Consequently, rework will inevitably be needed to eliminate the friction between the specification and reality. Even under the best of circumstances where talented people produce an outstanding specification document, at first experience with using their creation, they will be stimulated and start to perfect their original imperfect vision.

An alternative strategy is needed that better deals with the risk imposed on a project by the absence of supporting rigorous assumptions and is able to exploit the natural human tendency to evolve, rethink, and reorder one's needs based on experience.

2.3 PROTOTYPING DEFINITION STRATEGY

Application prototyping offers a viable and exciting alternative to prespecification. Prototyping is founded on the belief that people will better understand physical models than logical models and will naturally be able to suggest refinements. It exploits the human fallibilities of the user by placing them in a comfortable and enjoyable role: wise consumer. Prototyping sees the definition environment in a radically different perspective from the prespecification approach.

2.3.1   Underlying Assumptions of Prototyping Strategy

Figure 2.8 depicts the assumption/strategy pyramid that supports proto-typing. As would be expected, the foundation level assumptions are at the extreme opposite from those that support prespecification. The basic assumptions that make prototyping a "correct" strategy are as follows:

All requirements cannot be prespecified.

Quick build tools are available.

Inherent communication gap between project participants exists.

Active system model is required.

Rigorous approach is correct once requirements are known.

Extensive iteration is necessary, inevitable, desirable, and to be encouraged.

As was done with the assumptions that support prespecification, it is necessary to examine these assumptions more deeply.

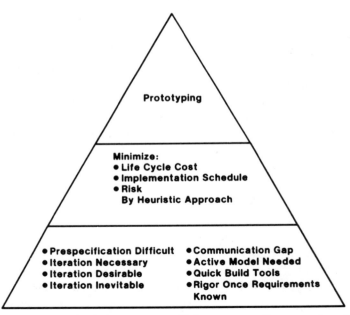

Figure 2.8 Assumption/strategy pyramid in support of prototyping. Realistic assumptions about the analysis environment support a prototyping strategy.

Assumption: All Requirements Cannot Be Prespecified

People find it extremely difficult to specify anything in detail. In practice, users are very good at stating the goals, objectives, and general directions they wish to move in, but are often unclear and undecided on exactly how they would like those things accomplished. Building a system is a gradual and continuing learning experience for all participants. How can one be asked to sign off on their final need when only 20% of the experience is concluded?

Data processing is one of the few disciplines that by deliberate policy does not deliver models. Physical models are seen as constraining on the designers and "hardware dependent." This may be true but so what? People test drive cars, walk through model houses, and even test out personal computer software before they buy it. Why not give the consumer the data flow diagrams for the check-balancing software and let them decide from that? People need help in determining their needs. The best help, as all other industries know, is a real world example. Examples can be studied and then critiqued in manageable units. Users cannot prespecify needs in detail because they don't know them in detail.

Conclusion. Specification is extremely difficult work and runs contrary to the normal life experience of most participants. The process is itself insidious. As people have each set of needs satisfied, the fulfillment of the needs spawns a new set. It is self-defeating to force incomplete ideas to be given the aura of detailed and final requirements.

Assumption: Quick Build Tools Are Available

It is only recently that this is true. Prototyping of large applications would not have been possible just 2 years ago. Today, the necessary software technology is coming into the marketplace that permits applications to be quickly modeled and even more importantly quickly changed. If the life cycle could be compressed into the definition stage and managed with malleable software, the cost-to-correct curve would no longer dictate our efforts.

The better tools that are available to perform prototyping are based on a few basic components:

Integrated Data Dictionary. A single repository for the definition and control of all system entities.

Flexible Data Base Management System. Provides for both ease of design to permit straightforward data modeling and ease of access to facilitate program development.

Nonprocedural Report Writer. Nonprocedural free form, parameter driven, and high default report writer that is thoroughly integrated with the dictionary.

Nonprocedural Query Language. Nonprocedural query language to permit ad hoc and saved queries. Again, thoroughly integrated with the dictionary.

Screen Generator. An interactive facility for painting "3270" type screens and performing automatic input editing such as numeric checks, table lookups, and so on.

Very High Level Language (VHLL). High function/high default procedural language for development of applications.

Automated Documentation. Since the dictionary serves as a record management system for the application, a dictionary report function is able to self-document the application.

Prototyper Work Bench. The tools are made available to the prototyper via an interactive work station providing ease of use and fully interactive development and results feedback.

This technology exists in various forms of development today and obsoletes the cost-to-correct curve. If "tinker-toy" prototypes can be built quickly, a "good guess" can be tested. If terribly wrong, it can be thrown away without incurring a large penalty. If it is in the right direction, it can be refined. Visions, concepts, ideas, and needs can all be tested in the "prototype laboratory" for correctness before a large commitment of resources is made.

Conclusion. Software technology is changing rapidly. We must not view our current options in terms of software that is 10 years old but rather view it in terms of what is available today. Software "tinker-toy technology" which is coming into existence provides a high degree of speed and flexibility.

Assumption: Communication Gap Between Project Participants

Communication is recognized as the fundamental problem of development. Even when people know what they want, it somehow comes out different in the translation. English is a poor specification language. Though certainly better, the prespecification tools lack the precision of any engineering specification language.

On the other hand, a user/prototyper discussion based on an operational set of screen dialogues is a very boolean experience. Discussion can take place in a simple question/answer manner:

Prototyper: The Customer Number is 6 position long, right justified, numeric, and must have a "3" in the 4th position—okay?

User: No, the "3" must be in the 5th position and I don't like the way you abbreviated Customer Number (CUS-NBR) on the screen. Could you make it CUST-NBR?

Prototyper: After you enter all the fields, hit the ENTER key. If any field is in error it will be highlighted in bright intensity and the message "Highlighted Fields Are Invalid" will appear on the bottom of the screen. Let me demonstrate.

User: This is unacceptable. My people will never understand computerese. In the message you must tell me the name of the field and exactly what is wrong.

Likewise all the project participants can massage the prototype in a boolean manner to test it from their unique perspective. Prototyping provides the best possible response to poor communication: a vivid and animated example.

Since logical specifications cannot provide such examples, and users have had trouble understanding the graphical diagramming techniques, a recent trend has been to suggest that users learn how to make and read the diagrams, that is, to buy a house you must be able to read the architects drawing. This idea as a response to poor communication is at best controversial. The user already has a demanding skill and profession. They have a critical day-to-day business function to perform. Does the comptroller really want or expect her line managers to be drawing flow diagrams or to be analyzing receivables?

Even if the users are willing, it must be recognized that flow diagrams are superficially easy to learn but upon examination are extremely complicated. This should not be surprising since they attempt to model highly complicated networks of processes and data. If it wasn't so complicated, there wouldn't be so many problems.

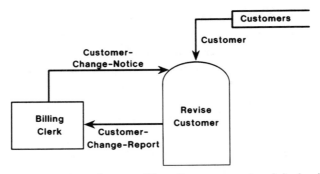

Figure 2.9   Data flow diagram. Flow diagrams are deceivingly simple.

Figure 2.9 is an example of a flow diagram. It, as a tree in a forest, is relatively simple to understand and discuss. It, however, is part of a big system, the forest. Figure 2.10 is the normalized data base structure that would be necessary to support a composite structured analysis technique in a mechanized fashion. It is obviously quite a complicated network of entities and relationships. Figure 2.9 can be read as follows:

> An external source, Billing Clerk, sends a Customer-Change-Notice data flow to a process "Revise Customer." To process the Customer-Change, access the data flow Customer from the Customer's file, apply the changes, and send a Customer-Change-Report data flow back to the Billing Clerk.

What could be easier for a user to understand?

Though Figure 2.9 as a tree is not all that complicated to comprehend, the user must also understand it in the perspective of Figure 2.10. To model a business environment the following entities are defined:

External Interface.   A source or receiver of system data flow outside the boundary of the system.

Element.   A basic unit of data.

Data Flow.   A collection of data elements in motion.

Process.   A set of transformation logic that is applied to data flows.

File.   A storage medium for data flows at rest.

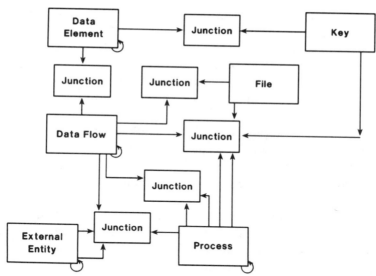

Figure 2.10   Data base required to support structured analysis. The records and relationships maintained by a data flow diagramming technique are quite complicated.

Key.   A collection of elements that define how a file is accessed.

These entities have multiple and complicated relationships with each other, for example:

A process receives data flows from external sources.

A process sends data flows to external sources.

A process sends data flows to another process.

A process receives data flows from another process.

A process reads data flows from a file with a key.

A process updates data flows to a file with a key.

A process may be a subprocess of another process.

A process can be decomposed into multiple subprocesses.

All the above happen at a given level of decomposition.

A level of decomposition must be balanced, that is, all the outputs must be derivable from the inputs, all the inputs and outputs must be derivable from a higher level.

When viewed in this manner, teaching users the definition technique is not all that attractive.

To understand the requirements proposal, both the forest and the trees need to be understood. Unfortunately, it is a circle. To understand the trees, you must understand the forest. To understand the forest, you must understand the trees. Perspective must be able to be interchanged rapidly to get a balanced understanding.

A menu driven prototype provides a familiar way to provide both tree/forest perspectives. As one descends the menu hierarchy, one descends into more and more detail. The consequences of having performed a detailed function can be understood by executing the effected function.

Conclusion.   The solution to the communication gap is not to try to make everybody a professional specifier but to permit everybody to receive specifications in a familiar medium. Working prototypes are a common sense means to accomplishing that.

Assumption: Active/Participatory Model of the System Is Required

Words, still pictures, and graphs are certainly good means of communication. However, they simply cannot effectively communicate motion. When the subject has living attributes, we need to see it animated to understand all the meanings and implications. Interactive systems require animated specifications. Users need to "play" system. It is one thing to look at hand drawn screens and pretend that there is motion but quite another to actually hit the enter key and watch the system happen before you. The difference is in orders of magnitude. There is no reason to settle for a passive specification if a dynamic one is possible.

Conclusion.   All parties will understand the implications of the proposal much better if it is animated. If a picture is worth a thousand words, an animated model is worth a thousand pictures.

Assumption: Rigorous Approach Is Correct Once Requirements Are Known

The endorsement of prototyping is not a license to abandon structured and disciplined building techniques where and when they are correct. For each part of the life cycle, a pragmatic assessment must be made to determine the best approach for that step. Most structured disciplines recommended the use of Inspections, a heuristic approach, to find errors in top down structured programs.

Heuristic approaches, like prototyping, work well in the early phases of development to help turn sand into cement. Rigorous concepts like span of control, coupling, and cohesion work well in the design and programming phases. One must be free to exercise professional judgment to select the best technique for each problem.

Conclusion.   The different phases of the life cycle have different needs. The early phases are better served by heuristic methods; the later phases are better served by discipline and structure.

Assumption: Extensive Iteration Is Inevitable, Necessary, and to be Encouraged

Users should be aggressively encouraged to change their mind. Improved insight should evolve from experience. Brainstorming of how to change a model when it sits before you is an exciting experience. People need a friendly environment, receptive to change, to maximize their potential contribution. Rigorous definition implicitly suppresses change after the definition stage since change implies poor analysis. You lock yourself into a far less than possible solution.

By encouraging iteration and experimentation through prototyping, you will deliver 2-year-old systems on cutover day. Figure 2.11 shows a system comfort curve. There are two comfort points for a user:

Minimum Comfort Point.   This is the minimum service the system must deliver for the user to live with the system.

Good Fit Comfort Point.   This is the point where the system's functionality as tested by operational reality meets most of the user's needs and the user is satisfied.

A good part of the postimplementation rework is targeted to first achieve the minimum comfort point (if not reached by a user/application sensitive time, the user will abandon the application as unworkable) and then achieve the good fit point at which the user actually likes using the sys-

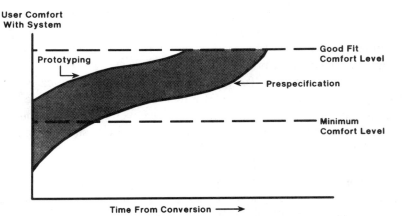

Figure 2.11 System comfort curve. A system must reach a minimum comfort point for the user to live with it.

tem. Since the user will get to experience the system by prototyping, the comfort curve will shift left: more of the "good fit" functions will be delivered on day one.

Conclusion.   Iteration is absolutely needed to develop ultimate needs. It permits a better fit to occur between the user and system earlier. Though you can certainly iterate within prespecification models, the iteration is constrained by the absence of animation. By discovering function while walking through the prototype, the whole system comfort curve can be shifted dramatically to the left resulting in a good fit between the user and system occurring sooner.

2.3.2   Prototyping Conclusion

Not surprisingly these assumptions lead to a different conclusion and overt strategy than prespecification. If you accept the prototyping assumptions as an accurate assessment of reality, then you will minimize life cycle cost, implementation schedule, and project risk by using a heuristic definition approach such as prototyping. Figure 2.12 depicts the SDLC with a prototyping strategy overlayed on it. Definition is exploded to permit user experience by creating a miniature life cycle within the definition phase. It is anticipated that experimentation will be necessary

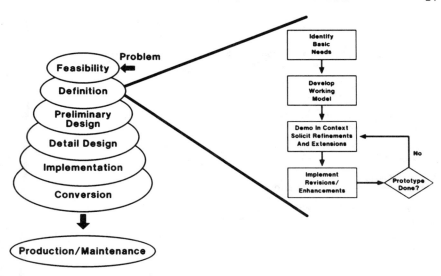

Figure 2.12   SDLC with prototyping. Definition is exploded to permit user experience by creating a miniature life cycle within the definition phase.

to discover the ultimate needs. The costly churning and looping, however, as occurred in Figure 2.7, is avoided by designed and planned iteration. When definition is now exited, all participants can have a high level of confidence that the product will be acceptable at conversion since they all experienced and refined it during definition.

2.4   PROTOTYPING VERSUS PRESPECIFICATION

The decision to choose a prespecification or prototyping approach is not one of comparing them directly but assessing the validity of the underlying supportive assumptions. Figure 2.13 provides a matrix summary of the assumptions for each strategy. Which set of assumptions do you find more appealing?

As a group, the assumptions that support "rigor" are highly suspect. We know from our experience that

- [ ] users have trouble prespecifying all their needs
- [ ] communication is always a problem
- [ ] a passive model leaves too much to the imagination

Assumption	Definition Strategy	
	Prespecification	Prototyping
1. Final And Complete Prespecification Possible	X	
Prespecification Extremely Difficult		X
2. Change To System Extremely Expensive	X	
Modern Quick Build Tools		X
3. Good Project Communication	X	
Inherent Communication Gap		X
4. Static Model Is Adequate	X	
Animated Model Required		X
5. Rigor Tautology	X	
Rigor Once Requirements Known		X
6. Iteration Proof Of Definition Failure	X	
Iteration Necessary, Desirable And Inevitable		X

Figure 2.13 Assumption/definition strategy matrix. Which set of assumptions is true?

As a group, the assumptions that support prototyping appeal to both experience and intuition. Practical experience indicates that they are often true. Figure 2.14 illustrates a typical engineering project life cycle. Notice the importance of prototyping to all phases. Most engineering disciplines routinely use prototyping to control risk and verify an idea before making major resource commitments. Prototyping eliminates the inherent project communication problems by permitting all debate to focus on physical anchors.

2.5   BENEFITS OF PROTOTYPING

The arguments to this point have centered on explaining why prototyping will often be a superior strategy. We will now turn our attention to itemizing both the qualitative and quantitative benefits that should occur as a result of using a prototyping approach.

2.5.1   Qualitative Benefits of Prototyping

By the adoption of a prototyping approach, the following qualitative benefits should be accrued.

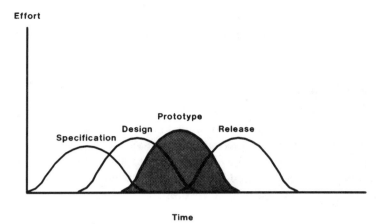

Figure 2.14 Engineering project life cycle. The building and evaluation of a prototype is the central step.

Benefit: Accommodate Decision-Making and Problem-Solving Styles of the Users

Users are not expected to become amateur specifiers to develop an application. To the contrary, they are merely asked to review and critique common media with which they are perfectly familiar. It is atypical to ask a user to approve a 6 level decomposition of a process. It is quite typical to ask them to enter data from a mock form to a screen and ask them how pleasant or irritating the experience was. The medium for communication accommodates the user, not the analyst.

Benefit: It Both Increases and Requires Active User Participation

The idea that the developer or project manager knows best is long dead. Users must be actively involved to insure the functionality and acceptability of the system. In conventional prespecification, users may or may not find reading the documents and attending the walk throughs exciting. They may sign off based on a careful analysis of the proposal or simply sign off to get rid of it.

Users can't wait to see a prototype. Their eyes light up and the ideas swirl as they experience the imperfect model. The users actively participate because they have a meaningful and comfortable medium to participate through. The problem becomes one of sorting out all the ideas and suggestions for tryout rather than an absence of innovative suggestions.

Benefit: It Provides a Vehicle for Validating Requirements

Question: How do you test the prototype to confirm that it works?

Answer: By attempting to perform the business service with the system. Either the model can perform each simulated business transaction or it can't. If it can't, either you have agreed not to mimic that function in the model or you have discovered input to your next iteration. Validation of a prototype is done by operational testing.

Benefit: It Provides a Facility to Permit Assessment of the Impact of the System on the Whole User Environment

Systems are not islands. They exist working in various levels of harmony with both manual and other mechanized systems. How disruptive will its introduction be? How well have the human interfaces been planned for? By executing the prototype, all parties can better assess the actual impact the new system will have on the entire work place.

Benefit: Permits Early Life Cycle Testing of Human/Machine Interfaces

At best, a working computer system is only half the battle. Systems consist of two major components: the personnel subsystem consisting of the activities done by people and the computer subsystem, consisting of activities performed by the machine. How well do they integrate? Should the interface point be moved to include more or less functionality? Is the system friendly or are the messages cryptic? Does it help and guide the user in performing a function or does it hinder the effort? If a machine response is required while a user is on the phone, can the needed screen be gotten promptly to or must six menus be accessed first? How will we operate if the system becomes inoperative? How will we restart when the system comes back up? If the goal is a paperless office, can the goal be realized? All these questions about how smooth the interface will be can be addressed much earlier with prototyping.

Benefit: It Provides Vivid Documentation to the Developer

Handoff between life cycle phases is a traditionally difficult problem. A new group of people have to be introduced to the vision. How do we communicate to them? Certainly we need to give them a great deal of what they customarily receive:

- ☐ elements descriptions
- ☐ record layouts
- ☐ screen layouts
- ☐ and so on

In addition, however, the idea can be demonstrated to them. Both the model and the hard copy documentation can communicate the need to the developer. Each one with its own strengths can serve to clarify the other.

You will also read later that it is suggested that a lead developer be part of the prototyping team. Just as the user is learning, so is the developer. The transfer should consequently go much smoother.

Benefit: It Is Behaviorally Feasible

In selecting a definition technique, it is important to recognize that human behavioral feasibility is an important aspect. Will people work in harmony with the method or is it only "theoretically" sound? People like to participate in prototyping projects. Building models is fast, fun, exciting, and extremely rewarding to all participants. It challenges them and brings out the best in creativity and innovation. It provides a true quality of work life experience since everybody can quickly see their contribution. Once a user department has had systems built by prototyping, they will not easily agree to revert to traditional approaches again.

Benefit: It Is an Additional Project Safety Valve

During the project feasibility stage, proposed systems are tested for technical, operational, and benefit feasibility. There is, however, another feasibility that needs testing: consensus feasibility. Systems that cross departments, organizations, and locations within a company will require compromise and agreement by a large range of diverse interests to operate successfully. Will each vested interest agree?

Prototyping provides a vehicle to test the consensus issue early. What if consensus is not achieved? Can system compromises be demonstrated to develop an agreement? If consensus cannot be reached, isn't it more cost beneficial to abandon the system at this point than after 2 more years of effort? Systems need to address political realities as well as operational ones. Prototypes can compromise conflicting needs.

Benefit: It Accommodates Uncertainty and Risk

Various studies have indicated that perhaps from 30–50% of all software development projects do not ever deliver an operational system or do so only after extensive modification. Medium- to large-scale application development is a high-risk undertaking. To minimize risk, models must be used to test the ideas. Only when the risk level has been negated by consensus acceptability of the model, are major resources expended to build the full system.

Benefit: Alleviates Project Communication Problem

Communication is the critical issue. If you have problems convincing people that a prototype is a superior medium for communicating, a simple experiment is suggested. In parallel, create both the traditional rigorous documentation and a prototype. First give people the rigorous documentation to read and solicit comments. Now present the prototype and again solicit comments. It has been my experience that the decision is no contest. Prototyping results in everybody having clear opinions on the system.

Benefit: Permits Both Forest and Tree Perspective

One of the strong concepts endorsed by rigorous techniques is decomposition, the ability to successively decompose a process (or other entity) into smaller pieces. This provides the ability to view things at different levels of detail and obtain both the forest and tree perspective of the system.

Prototyping accomplishes this same goal very effectively. Figure 2.15 shows a series of four screens that provide the desired top down perspective. Starting at the "home menu" (top left), we hit PF1 and descend to

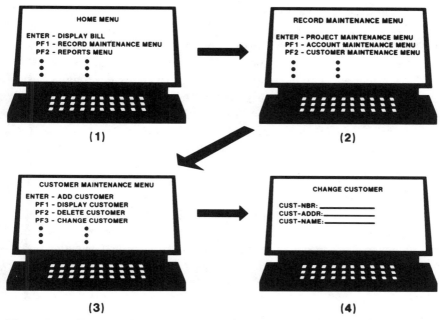

Figure 2.15   Forest and tree perspective. Prototyping permits the user to view the system at both perspectives.

the "record maintenance menu" (top right). We now hit PF2 and descend to the "customer record maintenance menu" (bottom left). We now hit PF3 and reach the detail "change customer" screen. Likewise other paths could be traversed to get a different perspective.

Desired top-down forest to tree presentation is available. The user, while studying the prototype, can either progress down hierarchically to experience ever more detail or can take a lateral slice and look at all the menus or functions at a given level.

Benefit: High Project Accountability and Visibility

The maximum time for delivering a first cut model should be 6–8 weeks for large systems and 2–3 weeks for small applications. As a consequence, there is high visibility as to "where are we." In the same manner, there is also clear accountability. The prototypers are responsible for delivering a working version and the user is accountable to review and critique it.

Benefit: Simple Project Management

As a consequence of the high visibility and accountability, it is not necessary to set in place elaborate project control systems with Gantt or Pert
diagrams. The iterations are shorter than the normal project report cycles
and a status report can be a demonstration. Project management consists
more of enabling the process by eliminating obstacles than methodically
checking progress. The short intervals between substantive deliverables,
the visibility of those deliverables, and the clear responsibility of each
party alleviates the need to impose sophisticated tracking mechanisms.

Benefit: Developer Apprenticeship Laboratory

Prototyping provides an excellent environment for inexperienced developers to gain extensive experience in a relatively short time. A junior
developer in a prototyping environment may work on 5–6 systems within
a year. From each experience, the apprentice will learn new technique
and appreciation of alternative system structures. Conventional development, naturally, requires much longer cycles per project. As a consequence, the less experienced developer's learning curve can be markedly
accelerated by placing him/her in a fast track environment.

Benefit: Systems by Example

After a while, a prototyping center will have developed a fairly wide range
of applications. The applications will demonstrate different solution approaches as well as a variety of function. All this of course can be demonstrated to users during the initial phase to

- [ ] show them what is possible
- [ ] stimulate their ideas
- [ ] let them identify styles they like or dislike
- [ ] permit them to explain their needs in terms of modifications to
  existing models

In essence, like an architect, the prototyper can exhibit a portfolio of
systems and see whether the client likes any. The hope is not to get an
exact match but to get a feel for the user's preferences. Later on during
discussions, the user will often say "could you do this like you did the
BROWSE in the *XYZ* system?"

This of course takes modeling to its logical end. Not only should we build custom models for the user, but also initiate the process by first showing and demonstrating existing models. With a good "system-by-example" portfolio, many previously confusing conversations can be reduced to "I want one just like that except add *ABC*."

Benefit: User Centered Definition

The entire thrust of this technique is to service the user. Data processing is a service organization. Its only reason for existence is to provide service to the other departments of the business. To do this, data processing must be responsive and helpful to users. Prototyping recognizes the user's preeminence in making the requirement decision but doesn't abandon them. The ability of a prototyper to work with the user at an evolving level of specificity and respond quickly provides the service that users want, need, and deserve.

Benefit: Interim Training Vehicle

At the conclusion of the prototyping cycle, there will exist the working image of the ultimate system. Though inappropriate to implement as a production system, it can serve as an interim training vehicle for the user while the actual system is being created.

Few other definition techniques can provide such a variety of benefits for the user, project manager, and developer. While qualitative in nature, they are clearly appealing to our experience and are consistent with our intuitive expectation with what would be expected to be accrued from a successful modeling technique.

2.5.2   Quantitative Benefits of Prototyping

Prototyping is not only an effective technique but it is also cost efficient. The following arguments demonstrate why cost efficiency is achieved.

Benefit: Trade Unplanned Rework for Planned Iteration

The life cycle model depicted in Figure 2.7 demonstrated the cost consequences of an ill-chosen definition strategy. The thrashing and the looping are expensive and since much of it occurs after conversion, it occurs at the highest premiums. The expense is primarily due to three factors:

1 The expense of the continual dismantling and reassembling of numerous software components, which is highly labor intensive.
2 The monstrous communication expense of coordinating the changes through numerous functional specialists.
3 The expense of continually redocumenting in full what is in actuality only interim and temporary documentation.

The developers have unintentionally delivered a prototype but are not prepared to refine it.

The model of the prototyping strategy as depicted in Figure 2.12 demonstrates how the costly thrashing and looping are controlled. By design, the costly loops are replaced by small manageable ones. When the actual development is inaugurated, most of the uncertainty that causes the thrashing has been eliminated. The savings are the result of trading unplanned rework for planned iteration.

Benefit: Left Justified System Comfort Curve

From one perspective, there are only two types of systems: those a user must use (such as an Accounts Payable), and those that they choose to use (Sales Trends). In the latter case it is very important to achieve a good fit on the system comfort curve (Figure 2.11) as soon as possible. While the Accounts Payable user will suffer through at a minimum acceptance level, there is no choice, the Sales Trend user will just throw the outputs in the garbage or not sign on if it is not a good fit. "Why be in pain, I'll do it my old way," the user will say. It is obvious that optional use systems accrue no cost benefit if they are not used. Consequently, the earlier the good fit, the greater the cost benefit achieved.

As previously discussed, prototyping helps discover those good fit features during requirements. At implementation, the user who "chooses to use" will have a system whose good fit point is justified left.

2.5.3 Cost Justification

Though these benefits are both quantitative in nature, they unfortunately lack a bottom line orientation, that is, neither tells you exactly how much you will save nor how to derive it. The problem here is the greater issue of software productivity measurement. There is as yet no consensus on how to measure productivity. If these quantitative arguments are unacceptable without actual figures and you work for an organization that insists on a bottom line justification, two techniques can be used that may serve adequately in providing surrogate cost justification measures:

Bebugging.   Deliberate bugging of software to discover error detection efficiency.

Function Point Measurement.   Measuring the cost of delivered system function.

If necessary, they offer an interesting and defendable approach to providing cost/benefit justification.

2.5.4   Bebugging

Bebugging is a method of determining the error correction efficiency of a process. It was introduced as a software metric technique by Tom Gibbs in his book "Software Metrics." The goal of bebugging is to plant representative errors in a product. Assuming that the errors are indeed representative of the normal set of errors, the efficiency with which they are detected, found planted errors divided by planted errors, should be indicative of the overall efficiency of the error detection process being used.

How does this relate to cost justifying prototyping? The basic reason for using prototyping is to develop the actual needs early in the life cycle. We would like to demonstrate that a prototype definition has a low level of errors. Since most errors that are detected late in the life cycle originate in incorrect requirements we would like to use bebugging of prototypes to demonstrate that prototype demonstrations are a high efficiency error detection process.

A method for performing this could be as follows:

1 Based on prior experience with rigorous definitions develop a representative set of errors. Sample errors may consist of
   a missing fields
   b incorrect field sizes

c   incorrect editing
d   missing function
e   incorrect function
f   extra function
g   function at illogical point
h   illogical system flow

2   Deliberately bebug a prototype.
3   During demonstrations to the user, keep careful records of the bugs detected.
4   Determine the error detection efficiency.
5   The computed error detection efficiency should be representative of the overall process.

This type of analysis could be used to make the following cost/benefit argument:

From the "cost-to-correct" and "source-of-system error" graphs, it is known that detecting and correcting errors early in the life cycle is a high productivity strategy. By bebugging prototypes, we have demonstrated an error detection rate of $X\%$. This compares most favorably with other techniques we have employed. Prototyping is consequently cost beneficial since it will relocate high cost correction to a lower cost part of the life cycle.

It is interesting to consider doing this in parallel with a prespecification document. Both the prototype and the structured specification could be bebugged. The results could provide an interesting set of statistics to prove the point.

2.5.5   Function Point Measurement

Function point measurement is a relatively new and novel approach to productivity measurement. The creator of it is Dr. Allan J. Albrecht of IBM. What is unique and special about it is that it completely avoids traditional measures like lines-of-code or error detection rates and instead concentrates only on measuring function delivered to the user. This makes a great deal of sense. It attempts to measure what the system does, not how it does it.

The basic thrust of the technique is to measure an application in terms of function delivered and the associated cost. Function consists of weighted summations of inputs, outputs, inquiries, master files, and complexity. The net result is a unit cost per function that is implementation technology independent.

This relates to cost justifying prototyping in the following way. You would like to demonstrate that systems which use prototyping have consistently lower cost/function than systems which under went prespecification. Assuming a parallel development, a system being done both ways concurrently, if measures were taken at the good fit point of the "system comfort curve" it would be expected that the prototype originated solution would be less costly due to the less costly iterations and earlier reaching of the good fit point.

It is not desirable to sidestep the question of cost justification and it is anticipated that both these suggested methods will be adequate if a clear cost justification is required. Unfortunately, until a clear productivity measure is adopted, any new approach will have to be judged primarily on its qualitative benefits and likelihood of cost savings.

2.6   IMPORTANCE OF PROTOTYPING

It is important that companies accrue the benefits that prototyping can provide. Productivity is the major issue of the 1980s and prototyping is an excellent method for improving life cycle productivity. Though there are literally hundreds of solutions offered to improve life cycle productivity, there really are only four primary strategies:

1   Data processing can offload work to software vendors.
2   Data processing can offload work to end users.
3   Data processing can trade expensive labor for intelligent software.
4   Data processing can work smarter.

Strategy: Offload Work to Vendors
The goal of this approach is to acquire finished applications from a commercial software vendor. Since many applications have well defined and common functionality and interface points, the vendor should be able to

deliver a solution at much less cost than attempting to develop internally the application from scratch.

Though this approach can certainly help, as a major pillar of a productivity strategy it will probably be unsuccessful because of the following:

1 Vendor provided software will only make up a small percent of your application portfolio.

2 Much of the application portfolio is composed of industry/product dependent software and customized intelligent software is a business edge.

3 You are an "exception" driven business. Most of your logic is "If then else." Though the vendor may provide you with a shell, all the hard work remains.

Currently, vendor provided software is being quite successful in helping the spread of personal computers by providing Information Center and decision support functions. As personal computers are made nodes in larger applications, they too will require application specific solutions.

Conclusion.   It will help, but, if you are a medium to a large size company probably not all that much for your Production Center applications.

Strategy: Offload Work to Users

The goal of this approach is to make everybody a programmer, just like self-dialing telephones made everyone an operator. Though this will certainly address the backlog of applications that are Information Center, decision support, or personal data base oriented, it will not really help in addressing the building of systems of any complexity.

There must be a clear distinction made between the types of applications end users will build and those that will in the foreseeable future have to still be built by professionals. As users attempt to move beyond retrieval and analysis applications, they will not be exempt from the same complexity and problems that the professional developers have had to address for years.

Conclusion.   This will certainly help in off loading the reporting and analysis types of applications.

Strategy: Trade Developer Labor for More Intelligent Software

The goal of this approach is to adapt more intelligent and functional software as development tools. Since the cost of executing instructions keeps declining while the cost of labor keeps rising, it is cost effective to let automated software function do as much work as possible and let the developer only have to develop the custom and exception procedure aspects of the application. This approach can pay extremely high productivity dividends.

Conclusion. It is not unreasonable to expect increases of developer productivity of 300–500% by using advanced software technology. However, regardless of software technology, it is necessary to build a user acceptable system.

Strategy: Work Smarter

As always, the key to making a major productivity improvement is to change horses. Any technology or technique as it goes through refinement will deliver smaller and smaller incremental gains. To achieve orders of magnitude increases, you must jump to a new solution space. Many of the best ideas for improving productivity like component engineering and data administration all fall into this category. Prototyping, likewise, is a work smarter solution. Perhaps eventually, artificial intelligent software will ask all the necessary questions and automatically deliver a solution. In the meantime, working smarter must equate with improving communication and prototyping does that extremely well.

Conclusion. Working smarter is the best means to achieve dramatic productivity improvement.

Prototyping, as a work smarter approach, offers productivity benefits too important to ignore. If you accept the original thesis that requirements definition is the critical step to life cycle effectiveness, it is imperative to adopt definition strategies that will directly solve the definition problem.

2.7 SPECTRUM OF DEFINITION ENVIRONMENTS

This chapter has heavily emphasized the benefits of prototyping and in doing so has been extremely negative on prespecification approaches.

Figure 2.16 Spectrum of definition environments. Definition environments vary widely in the ability of the participants to specify in detail.

This was necessary to make a strong differentiation without constantly qualifying every sentence. The day-to-day reality, of course, is not so black and white.

There is a wide spectrum of definition environments and situations (Figure 2.16) under which requirements definition must take place. At one extreme, a well understood conceptual model of the problem exists. The users are able and willing to specify in great detail. The analyst and developers are familiar with the application and are able to elicit, evaluate, and integrate the needs effectively. A management consensus has been reached on the boundaries, functions, and flavor of the system. A small group of users is able to work intimately and continually with the project team to provide good communication.

At the other extreme, no conceptual model of the problem has been formulated. Users have some goals and objectives but have not been able to translate them beyond the "fuzzy" level of detail. Users are spread across departments and a consensus on the functionality of the system and each department's role does not clearly exist. The developers and analysts are unfamiliar with the application.

In the first case, there is nominal risk or uncertainty. To proceed with a rigorous approach may work very well. In the later case, there is a high degree of risk and uncertainty. Prototyping is necessary just to provide an anchor point from which to start meaningful discussion. Such extremes in environmental conditions require different approaches.

In practice, few problems are at the extreme points, neither is the situation perfect nor is it hopelessly disorganized. Most problems have characteristics of both and require prudent mixing (Figure 2.17) to derive a project by project dependent optimum approach. What is critically important is the availability of both techniques to address properly the environmental conditions that are true about a particular project.

Currently, organizations automatically assume that prespecification should be applied without judging the situation. To the contrary, most conventional business applications have conditions that make prototyping more appropriate. Nevertheless, there will always remain parts of projects or entire projects where the conditions are such that prespecification will prove superior.

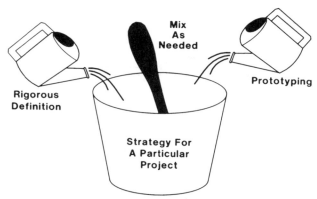

Figure 2.17 Composite definition strategy. Both prototyping and prespecification need to be applied on a project-by-project sensitive basis.

2.8 BUSINESS CASE CONCLUSION

The business case that supports prototyping is too strong to ignore. Given the unpleasant state of software development, the benefits that prototyping can deliver can play a major part in revitalizing the entire development process. Even the most skeptical management should not permit such a profitable opportunity to be left untested.

The key to the success of the prototyping is that it focuses on the user. Almost all the benefits directly involve the user. User centered analysis is the means to derive requirements that will be ultimately acceptable. Prototyping can accomplish this by the natural and pleasant manner it permits the user to participate actively in the definition process.

At a more abstract level, prototyping is part of the continuing trend in data processing to more friendlier and intimate solutions (Figure 2.18). Just as programming languages have progressed from cryptic machine language to high level program generators and the computer has moved from the distant batch machine room to the desk top, requirements definition is evolving toward prototyping. The prior definition techniques were only interim and temporary steps until animated definition by prototyping was technologically possible.

The building of application models is a no-lose situation. At best, following a few quick iterations, a consensus is achieved and the actual development can commence with a high level of confidence of success. At worse, the prototype makes visible all the issues requiring resolution

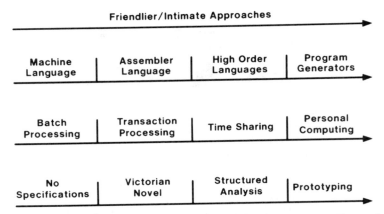

Figure 2.18 Trends in data processing. A constant thrust to friendlier and more intimate solutions.

before the construction of the actual system can begin. In both cases, dear development dollars and resources are not expended until project risk has been constrained.

The development of requirements is an imperfect process done by imperfect people. It is both an expensive and high-risk endeavor. Prototyping dramatically reduces the exposure to failure. There is little more one can expect from a definition technique.

THREE

THE PROTOTYPE
LIFE CYCLE

The purpose of this chapter is to provide a methodology for performing requirements definition by prototyping within the framework of an SDLC. Rapidly building large system models is a challenging task and requires structure for the prototypers so that they may present the user with both a malleable and responsive interface to the model. The primary thrust of the methodology is to perform logical prototyping. The technique emphasizes the rapid development of a meaningful model from the user's perspective without concern for the operational constraints of the system. A specific process in the latter part of the methodology is used to identify those details of the application that require rigorous specification, that is, record volumes, transaction rates, system reliability, and so on. The prototype is viewed to be exactly what its name implies, a model, and it is not intended to be implemented directly without all the necessary remaining work to create a complete system.

The following topics are developed to explain the prototype life cycle:

The Need for a Life Cycle. Explains why a life cycle approach is needed to support the prototyping of nontrivial applications.

The Prototype Life Cycle.   Explains the activities that are conducted in each phase of the life cycle.

Prototyping Principles and Tactics.   Explains certain concepts and methods that enable the prototyping process.

Hybrid Prototyping Strategies.   Explains alternative techniques for performing prototyping.

Implement the Prototype.   Analyzes the issues surrounding the question of whether the prototype can and should be implemented as a production system at the end of the prototype life cycle.

A New System Development Life Cycle.   Proposes a new view of the full development life cycle as a result of adapting a prototyping methodology.

3.1   THE NEED FOR A LIFE CYCLE

It may appear somewhat surprising and contradictory for a person who strongly endorses prototyping to be proposing a structured set of procedures to define the prototyping task. After all, "Didn't you say that all that detail and rigor was part of the problem?" An important distinction must be made between a "good guess" and implementing that "good guess." The user is being given a flexible, familiar, and malleable medium to work within. The prototyper still has to build with elements, records, screens, reports, programs, and so on, a working system (no one tells the software that it is only a prototype). To accomplish this physical act, the prototyper, as any other application builder, needs guidelines and rules to permit efficient development. If the distinction between the user's view and the prototyper's role is not made, we will have to expect either a miracle or magic to deliver a working and refinable model.

What is special about prototyping as opposed to other techniques is the user's perception and view of the process. The prototyper still has to build a working model with all the problems inherent in building any computer application. Software bugs do not make exceptions for prototypers. To build a prototype that will exhibit substantive functionality and be changeable as iteration proceeds requires order and structure for the builders.

Figure 3.1 is a stereotype portrayal of a prototyping session. The image

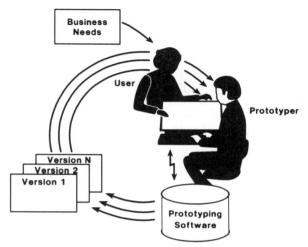

Figure 3.1 Stereotype prototyping session. The impression is often given that prototyping results in an instant system.

and impression is made of a user and prototyper talking over the application at the terminal and then quickly a screen is generated. The user and prototyper discuss the results and within a matter of minutes the application is finished. The user walks away contented and the prototyper, after a few finishing touches, arranges to put the application into production.

This image is misleading for a number of reasons:

It gives the impression that building prototypes is a trivial matter.

It gives the impression that only small applications are suitable for prototyping.

It gives the impression that the functionality delivered by a prototype must be extremely limited.

It gives the impression that the user has a simple task.

None of the following impressions are true.

Impression: Building Prototypes Is a Trivial Matter

Delivering an effective prototype is a demanding exercise. The benefits previously discussed are not the result of a casual undertaking. Building a system, whether a real one with finished detail work or a prototype with imperfect function and rough edges, is an exacting task. It must work, realistically approximate the actual system's functionality, be ergonomically sound, and be documented. In choosing to build a prototype, only a few burdens are eliminated: performance, optimum access strategies, each and every function, and so on. However, absolution from the user's view is minimal. If the prototype is truly to be the archetype of the target system, permit the user a meaningful experience, and the developer vivid documentation, it must be a truly representative model.

The image portrayed in Figure 3.1 is appropriate in terms of rapid refinement during the iteration phases. The initial building and early iterations where a throwaway situation is possible should not be viewed as "instant systems." In many ways, the building of models requires more planning and care than building a real system. The prototypers are committed to rapid revision, a complex task. Being accountable for revising a model of 150 data elements, 25 screens, and 55 programs that was casually assembled is an impossible task.

Impression: Prototypes Are Only for Small Applications

To be of value to a development organization, a requirements methodology must have utility in addressing the more difficult applications. Defining small systems, even under an organized chaos approach, can work reasonably well.

The impression given by Figure 3.1 of an instant prototype conveys a message to professional system builders that this approach must only apply to small applications. After all, it is known from experience that building systems of any size and complexity is not an instant effort even if most operational considerations are ignored.

Prototyping, when properly managed can deliver high function models of large applications in a relatively short time. To be of value, an initial model should be sufficient in size to permit both forest and detail perspective. This is not met by the delivery of only three screens of an application with potentially 30 screens. Though not instant, relative to traditional development, the delivery of an operational model with 30 screens in 6 weeks is literally overnight. There is no reason why prototyping should be limited to small applications and to be truly beneficial must be used on the complex and bigger problems.

Impression: Prototype Functionality Is Limited

The whole essence of a model is that it personifies the character, tone, and presentation of the actual. A model that will be shown to be a liar will yield little benefit. In addition to the accusations of deception, the user will merely start to iterate with the delivered system since a prototype of a different system was received at implementation. To be of value, the prototype must accurately demonstrate

- all primary functionality
- error detection and correction procedures
- interscreen flow
- human/machine interface

If the model fails to represent accurately these characteristics or they are radically altered in final delivery, two systems, not one, were prototyped.

Figure 3.1, by the impression of an "instant system," results in the professional system builder dismissing the functionality that can be delivered by a model. If production system function such as

- "tunnel" field editing
- cross field editing
- cross record editing
- duplicate checking
- error correction cycles
- ADD, CHANGE, DELETE, MODIFY functionality

are to be included, that simply will not be the result of an instant process.

For a prototype to be effective it must deliver a miniature in size but facsimile in function of the ultimate application. To start out with less of a goal is a waste of time and resources. Consequently, a life cycle is needed to permit the building of high function models. Without a defined process to follow as a road map, a meaningful prototype would be impossible to deliver.

Impression: The User Has a Simple Task

The user has a highly responsible and critical task. For the first time in system development, the user is actually being given the opportunity to

take control of her own destiny. The prototypers will organize and structure an initial vision, but the user must bless that vision. It is quite different to concur with an operational model than a paper model. The ambiguity is gone.

The user must also actively and fully participate in the process. This
means judging the usefulness of every screen, report, and human interface procedure. This is hard work. Though a quick discussion at the
terminal will refine a piece of the system, the user is being asked to
endorse the entire application.

This is as it should be. In the end, the user must judge and accept the
effectiveness of the application. Who else is capable? Along with this
status comes the associated responsibility and accountability for the
product.

The question posed was: "Why does prototyping require a life cycle?"
A life cycle is necessary to insure the delivery of a malleable prototype
with sufficient functionality and completeness to be representative of the
ultimate system. A prototype without these attributes will result in the
same thrashing and churning that is characteristic of traditional analysis
methods and the anticipated benefits will be lost.

3.2   THE PROTOTYPE LIFE CYCLE

Figure 3.2 is an illustration of the prototyping life cycle. It is the complete
model of the life cycle that was presented in a simpler form in the introduction (Figure 1.2). Three major steps that are required for delivering
a complete requirements definition have been added.

1   A candidacy selection step to determine the suitability of a system
    for prototyping.
2   A step to perform specification of components requiring detail
    definition.
3   A clean up step to complete any prototype loose ends and insure
    sufficient documentation for handoff to the next phase.

The earlier model, which was presented in Chapter 1, is adequate for
introducing the concept but is insufficient to deliver a finished specification that can serve as the basis for all the remaining development tasks.

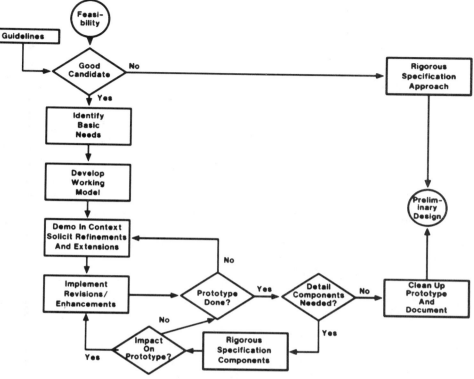

Figure 3.2 Requirements definition by prototyping. Prototyping provides a comprehensive approach to defining all the application requirements.

As illustrated, the prototype life cycle is bounded by other phases of the SDLC. The completion of a feasibility study should normally precede the initiation of requirements definition by prototyping. Much of the typical items documented in a feasibility study such as

- ☐ review of the existing business environment
- ☐ identification of current operational problems
- ☐ business goals, objectives, and opportunities to be addressed by the application
- ☐ major constraints on the application
- ☐ system boundaries and interface points
- ☐ schedule of user organizations and project representatives

- ☐ macroview of system inputs, functions, outputs
- ☐ cost/benefit goals
- ☐ application position in overall system plan

can be a great asset to the prototyping team in getting acculturated to the users environment. If a feasibility step was bypassed, much of this information will need to be obtained during the "identify user's needs" step. To avoid the delay, it is advantageous that feasibility should be done as a distinct step.

Bounding the prototyping life cycle on the exit side is preliminary design (PD). In a prespecification environment, PD initiates the process of defining a physical solution. With a prototyping approach, PD serves the function of analyzing and dissecting the prototype to determine

- ☐ optimum access paths
- ☐ logical data views
- ☐ reusable functions
- ☐ sizing
- ☐ actual physical data base design

It is not intended that the prototype should be directly implemented. This is necessary since much of what is required of a real production system is not available:

- ☐ performance engineering
- ☐ recovery procedures
- ☐ operational documentation
- ☐ implementation guides
- ☐ conversion aids
- ☐ quality control testing

The issue of whether the actual application should be developed by enhancing the prototype or using it as a requirements document to serve as the reference for the new system will be discussed fully in Section 3.5. The point, however, is simple. Unless you are dealing with a trivial application, how can you implement that which is untested, undocumented (from a user and operational perspective), and built without regard for performance? The prototype is a model. The remaining phases of the life cycle are required to fully complete the application.

The prototype life cycle consists of 10 steps. The goal of the process is to deliver a requirements definition that meets both the user's needs and the developer's needs. For each life cycle step, a statement of purpose, description of the step activities, and the exit criteria will be provided. This life cycle is targeted for building large system models. Hybrid strategies that address less complex situations will be discussed in section 3.4.

3.2.1 Good Candidate

Purpose

The purpose of this step is to determine whether the application is a suitable project for prototyping or whether the use of prespecification would be more appropriate.

Description

The appropriateness of prototyping as the primary definition strategy for a project needs to be determined. Just as automatically assuming that prespecification is correct has been argued to be inappropriate, the same is true in robotically building models of all and every new system without considered judgment. Suitability requires a balanced evaluation of a number of factors: system structure, logic structure, user characteristics, project management, application constraints, and environmental assumptions.

Candidacy Factor: System Structure

All system structures are not good candidates. Applications that are extensively batch in nature are inappropriate candidate systems. Structures such as batch edit/update and batch transform do not lend themselves to efficient prototyping. Applications that are full-screen terminal oriented (data base transaction processing) with periodic data base reporting and interfacing are good candidates.

Interactive applications lend themselves well to prototyping. User participation is easily encouraged. The essence of prototyping, developing the user/machine interface, does not apply to batch systems. Batch systems do not create a conducive environment for rapid iteration.

Additionally, software technology is much more advanced in supporting the rapid development of on-line systems than batch. This is because, in many ways, the on-line model is simpler. A transaction tends to be highly functional within a well defined set of boundaries. Batch programs tend to be a Pandora's box. They tend to have to do a great deal of supplementary functions to support the primary purpose.

Conclusion. A good system structure then is as follows: The application should be on-line/transaction processing oriented.

Candidacy Factor: Logic Structure

Applications that represent operational support systems, record management systems, or structured management information systems are good candidates. Applications that are heavily algorithmic, decision support, or ad hoc retrieval and analysis are poor candidates. Good candidate systems are those that "are the business" rather than "about the business."

Prototypes are appropriate for structured systems. The previously mentioned good candidates, record management, operational support, and management information systems, normally have a large structured component. To the contrary, decision support and ad hoc retrieval applications are by nature nonpredictive. The essence of these applications is the creation of a comprehensive data base and the selection of appropriate end-user software facilities. The Information Center, personal computing, and end-user software is the appropriate solution to these types of problems.

Prototypes are inappropriate for algorithmic based problems. Applications with few data elements and records but thousands of lines of algorithms do not lend themselves to leverage. The application is the labor intensive implementation of the algorithm. To be a good prototype candidate, the application should emphasize record maintenance and manipulation as opposed to extensive procedural logic.

Conclusion. A good logic structure then is as follows: The application should be a structured problem with a large amount of data elements and record relationships but a small amount of algorithmic processes.

Candidacy Factor: User Characteristics

As early as possible, the user should receive a presentation on the prototyping life cycle. The user must be informed as to the difference between prototyping and prespecification. The importance and responsibility of the user's role must be emphasized. A user who will be a good candidate will have the following characteristics:

The user will be dissatisfied with prespecification experiences.

The user will be willing and able to devote resources to defining and revising the model.

The user will be uncertain about detail requirements.

The user will be able and willing to accept decision-making responsibility.

All definition techniques require user time, interest, and participation. Prototyping, however, is especially unique. When a model is ready to be reviewed, the process comes to a loud and visible halt if nobody will review it. Users cannot abrogate their responsibility. There is not a 60 pound paper document to sign off on whether it was read or not. Decisions have to be made about the ultimate acceptability of each screen. Some users will find this responsibility welcome but they will be apprehensive as well. They should be aware of how they interface to the prototyping process before they start it.

Conclusion. A good user is one who is willing, able, and ready to actively participate.

Candidacy Factor: Application Constraints

A system that must be operational yesterday is a poor prototyping candidate. Such systems require a "good guess" implementation, not a "good guess" model. Consideration and iteration of a model take time. All parties must be able to let the requirements develop as they do. Artificial time constraints will result in an incomplete model, serving as a poor definition to the developers.

Conclusion. Prototyping is inappropriate for crash projects.

Candidacy Factor: Project Management

The project manager like the user must feel comfortable with the process and should be educated in it fully. A project manager who by experience or inclination is heavily committed to prespecification will find prototyping alarming and be a negative influence on the process. Especially important to the prototypers is the project manager's control of the project. Has a feasibility report been done? Will the manager help provide subject matter experts?

Conclusion. The project manager must be willing to work with the method.

Candidacy Factor: Project Environment

The key environmental assumptions that dictate a rigorous or prototyping approach need to be assessed:

Is meaningful prespecification possible?

Is good communication possible?

Is a descriptive/graphical model adequate?

These are the barometers that dictate which technique is more appropriate.

Conclusion. The primary definition technique should be selected based on each project's needs.

The selection of good candidate systems for prototyping is a project sensitive one. All factors should be considered to reach a prudent judgment. In practice, this is not as formal as may be implied by the preceding. Except when structurally inappropriate, most users will want to work with a prototype. In addition, the environmental factors will almost always tend to support prototyping. The candidacy question can often be phrased as follows: "Why shouldn't we use prototyping in this situation?"

Exit Criteria

This step may be exited when consideration of all factors is sufficient to decide whether prototyping or prespecification is an appropriate strategy.

3.2.2 Identify Business Needs

Purpose

The purpose of this step is to develop sufficient understanding of the business problem to enable the design and construction of the initial model.

Description

The first step in any requirements approach is needs analysis. What is particularly important is the identification of the application's "basic" needs. For most applications, there is a core set of data and function that compose its heart. Most of the application is ancillary to this heart and is derivable from it. One should aim to identify the key items that form the nucleus for the system.

If a feasibility report preceded this step, a great deal of valuable information will be immediately available. Of particular utility to the prototypers are the system goals, objectives, business opportunities, and problems. Embodied in these narratives are the forest vision for the application. They provide a target to aim for. Though details will tend to vary, these items tend to stay stable. Since these items serve as an important information source, if the study report was skipped, they will have to be developed as part of this step.

There are no tricks to discovering requirements. Current systems have to be reviewed, users interviewed, business directives researched, and so on. Discovering requirements is hard work. There are, however, two major differences between needs analysis in this context and traditional methods—predictive modeling and "good guess" exist.

Predictive modeling refers to a set of principles that allege that applications are highly redundant. The basic philosophy of predictive modeling is that for conventional business applications, a great deal of the requirements are known simply by knowing the nature of the application. Applications tend to be very similar. Though the application jargon and peculiarities tend to camouflage the commonality, most experienced system builders will readily admit that "they're all the same." As a consequence, applications can be viewed in terms of system models, data models, function models, design models, edit models, and report models. Analysis, as a result of this "model" philosophy, tends to center on matching the problem to the appropriate junction of models and identifying the "exception" characteristics of the problem.

This is quite different from other methods that start out with a blank sheet and need to have every detail documented. A complete data flow representing every detail of an ADD or DELETE function is hardly called for. Rather, having experience in doing ADD-type functions in multiple other applications, the question really is "Do you do anything special when you ADD an *ABC*?" The prototyper, of course, could explain to the user the model ADD logic structure. The user could then explain how, if at all, the situation is different.

When one views analysis in terms of matching models, a key variable to be determined is the data. What are the elements, record groupings, and inter-record relationships? When applications are viewed in terms of models, in a great many instances, knowledge of the data base structure equates to understanding the basic needs of the application. To analyze the data, a methodology is required. The best available technique is third normal form analysis. It will be discussed in Section 3.3.2.

The other major difference between traditional analysis and analysis in this context is that it does not have to be complete nor perfect, only a "good guess." The goal of a prespecification document is a complete, consistent, and perfect statement of requirements. The goal of needs analysis here is to gather sufficient information to initiate the model. Only enough has to be gathered so that by matching the variables with the models that a first cut prototype can be constructed.

During analysis, it is advantageous to let the user browse the portfolio of prior built applications. This is referred to as system by example. The user is able to become acquainted with the established product line. As a consequence, requirements can be couched in terms of the example systems. Similarly, the prototyper can show possible implementations to the user to insure mutual understanding.

During the analysis step, a great deal of information will be collected. The element definitions, function descriptions, and so on, all need to be stored in a recallable and manipulative medium. The best selection is a data dictionary. An interactive dictionary must be used to store the gathered information. It is the most efficient way to analyze and manipulate it and positions the prototypers for the building process without the need to reenter the information into machine readable format later.

Prototyping does not exonerate the participants from attempting to determine requirements up front. The user, project manager, and prototyper must meet and through intensive sessions develop sufficient detail to permit a meaningful first cut effort. Throw away situations, though acknowledged and accepted as part of the technique, are certainly not desirable and should be avoided. The only way to avoid unnecessary throwaways is to devote enough time to analysis to enable a "good guess." Predictive modeling minimizes the effort but there is no magic. Iteration is not a license to do careless and sloppy work.

The initial determination of user needs is a key step to the success of the life cycle. An initial model that is below 60% accurate will be disappointing and dampen user enthusiasm for the method. Iteration is used to refine and complete the vision, not to define it completely. The use of prototyping should not serve as an excuse to avoid meaningful analysis prior to construction of the first model.

Exit Criteria

This step may be exited when the prototypers are able to explain to the other project participants the business application at a reasonable level of understanding. If you can't explain it, how can you build it or create a conceptual solution?

3.2.3 Develop Working Model

Purpose

The purpose of this step is to build the initial version of the prototype.

Description

A working model must be delivered with sufficient depth and breadth to permit meaningful discussion and iteration to begin. It is more important to deliver a variety of functions with dull screens than perfect ergonomic masterpieces with little function behind them. The initial thrust has to be to demonstrate the completeness of the understanding of the business problem. Screens and reports can be human engineered during iteration. Content, not presentation, should be the primary goal of the initial model.

There is a logical order for building a model. The structure of the data base fairly well dictates the order of building subsystems. Within a subsystem, a typical order would be to build ADD, DISPLAY, DELETE, and MODIFY, then any other functions. The reasons for this are obvious, they are the primitive necessities of system life. If nothing else is delivered, the minimum necessities must be to permit a demonstration. If you can't do the four basic functions, there will be little to demonstrate or talk about. Periodic batch reporting and interfacing can normally be delayed till later cycles. If the data base is defined correctly, both these functions should be straightforward.

The building of the model should be "value added." There are "correct" structures that repeatedly need to be included in applications. Proper design items such as

- ☐   date/time stamping of records
- ☐   control totals
- ☐   audit trails
- ☐   standard screen headings and footings

should be provided as appropriate whether explicitly requested or not. All disciplines have proven constructs that when intelligently applied over time prove themselves to deliver consistently better products. The quality of the subject application should start with the best possible infrastructure from the beginning. This is not meant to imply that each item must be delivered in the initial version. Only that the shell should allow for them as and when iteration requires.

Since an application has to be designed and that design communicated, the question arises: How do you do that quickly? The answer is to use flexible communication media such as blackboards and paper easels. A prototyping work room requires wall-to-wall blackboards. Typical design items that would be placed on the blackboards are

- ☐   a data base design diagram
- ☐   names of system records
- ☐   structure chart of on-line programs
- ☐   names of programs and screens

Blackboards are a very receptive medium not only for communication but also for iteration. They are efficiently erasable and revisable.

Standards are very important to a prototyping team. Without them, everything is a personal product without availability to other team members. Communication between team members requires adhering to a productive set of rules.

In building the model, it is important to take advantage of the primary prototyping absolution; it only has to work in miniature. The design does not need to consider performance, optimum access, restartability, and so on. Only the user's view of the problem needs to be considered. This greatly simplifies the design effort.

The quality of the initial model is important to the success of the remainder of the life cycle. If it is grossly incomplete, it will serve as a poor anchor. If it takes too long to deliver due to completeness, it will not be responsive and a possible "bad guess" will need a large amount of refinement. If the model concentrates on delivering the "heart" of the application, iteration will start with a model that is an excellent anchor.

The time required to deliver an initial model will vary in proportion to its size, complexity, and completeness. To be effective, a target delivery of 3–6 weeks would appear highly desirable. This is enough time to both develop meaningful function and retain the user's interest. A maximum time period would be 2 months. After 2 months, a system, not a model, is being delivered.

A prototyping team should consist of two people with a swing person in a support role. If you are to achieve unity in design and permit rapid communication, a small team must suffice. A lead prototyper in this arrangement can design a unified vision and control its implementation. A large prototyping team results in the same communication problems and need for formalized procedures as any other development effort. Speed, flexibility, and unity of purpose requires smallness.

Exit Criteria

This step may be exited when a prototype of the application works and is sufficient in content to permit meaningful discussion.

3.2.4 Demonstrate in Context-Solicit Refinements and Extensions

Purpose

The purpose of this step is to develop new and revised requirements as a result of having all necessary people observe, critique, and experience the model.

Description

This step provides the opportunity to evolve the vision and functionality of the system. The model, as a baseline, is both demonstrated to and operated by the users. For the first time, the development process permits people to be human. They can evolve their understanding and change their mind without penalty.

The prototypers should aggressively prompt all reviewers for refinement. The model should be explained fully with the rationale for what it does. It, however, should not be defended. It is intended to be a moving anchor point.

Early iterations should concentrate on

- ☐ macro acceptance of the thrust of the model by the user. Are we in the ballpark? Does it look familiar?
- ☐ detection of gross oversights. Did we miss a record type?
- ☐ user familiarity and comfort in operating the model

Later iterations would tend to concentrate on

- ☐ discovering missing or incorrect function
- ☐ testing ideas and suggestions
- ☐ improving the user/system interface

As would be expected, the iterations move from gross consideration of the model to fine tuning.

A structured review process should be used to analyze the model in-depth. A member of the team may act as scribe to record changes and suggestions as the system is walked through. At the completion of each review session, the scribe can restate the agreed to changes to insure accuracy. Of course, whenever possible, the changes should be made directly and immediately in front of the reviewer for confirmation and acceptance.

Analyzing a prototype is hard work. The user is being given the opportunity to "test drive" the new system. Each screen, function, edit, error message, and so on, needs to be reviewed for acceptability. That which is not carefully analyzed now will not yield system life cycle benefits. The prototype should not be viewed as a toy but a serious exercise requiring careful and considered analysis. The simple and recurring question that should be asked of the user is: "Is this acceptable?"

It is important that the prototypers do not associate their success with the delivery of perfect models. If the process works correctly, users will in fact request significant revisions. This is not a criticism of the proto-typer but a natural result of the process. The goal of the prototype is to stimulate innovation and creativity, not to protect a "guess."

As the prototype takes fuller shape, its validity should be tested by "playacting" the business environment. If possible, the placement of a terminal in the user's facilities where a controlled parallel processing test could be conducted is an excellent way to test the model's ability to serve the user under actual conditions.

The model should be reviewed by all relevant parties. This might in-clude a vice-president who is funding it as well as the line user who will actually operate it. Everybody has a different perspective of what they want and expect from the application. When it is least painful, this is the time to obtain the needed consensus.

Exit Criteria

This step may be exited when a revised set of requirements is determined from the demonstration process.

3.2.5   Implement Revisions/Enhancements

Purpose

The purpose of this step is to bring the prototype into harmony with the user's revised expectations.

Description

As a consequence of the prior step, it is likely that revised functionality has been requested. In the event that a major misunderstanding has oc-curred or experiencing the application radically changes the user's desires, a throwaway situation could occur. This is not as ominous as it sounds. Unless a complete failure of communication occurred, it is likely that many if not all of the raw pieces of the prototype could be salvaged and serve as the basis for the new model. Throwaways, though unpleasant, are part of the discovery process. Much better to throw it away now, than after 3 years of effort.

In the more likely situation, the existing model will be incrementally changed. This requires the ability to control the positive and negative effects of rippling. This dictates that not only must a dictionary have been used for defining the application but also the dictionary must be recording all the relationships between the system components. If the users request that an edit be revised on a field, all the places where the edit occurs need to be identified. If a mechanized record of the use of each system part does not exist, the revision process will be error prone and time consuming. The availability of an active integrated dictionary that manages the development process is consequently a key software requirement for the prototyping software.

The revision process should occur as quickly as possible to maintain project momentum and user interest. Early iterations may take as long as a week, later iterations a few minutes (in front of the user). At the completion of this step, the model will be ready for another demonstration. All changes and their impact on corollary functions must be demonstrated. Requested changes are not always as desirable upon witnessing them as had been anticipated.

In certain cases, especially if the user is actively experimenting, it is advantageous to maintain both the before and after versions of the model. Not only does this provide an easy fallback if the user reverts, but showing both alternatives concurrently is an extremely powerful way to help make decisions.

Exit Criteria

This step may be exited when the agreed to revisions have been added to the prototype.

3.2.6   Prototype Done

Purpose

The purpose of this step is to determine whether the substance of the application has been captured and the iteration cycle may be concluded.

Description

It is to be expected that each successive refinement will start to deliver more marginal improvement to the model. Changes will migrate from examining function to evaluating the human interface. As each iteration starts to deliver more marginal improvement, a consensus will be reached by the reviewers that they are satisfied: "I want that!" The model does in fact illustrate the function and presentation of the ultimate system.

This event is readily observable. When the only requests for changes are to correct the spelling on an error message or move a literal one column on a screen, it is obvious that we are dealing with the most minor irritants. This is radically different from dealing with an early iteration when additional screens and functions are being requested.

Exit Criteria

This step is exited to the "demonstration" step if the prototype must still undergo refinement. This step is exited to the "detail components needed" step if the model has achieved consensus approval.

3.2.7 Detail Components Needed

Purpose

The purpose of this step is to determine whether any definition deliverables are missing that require rigorous specification.

Description

Nontrivial applications have definition components that have always and in the foreseeable future will always require detailed specification. Though certain items may be ignored to simplify and optimize the development of a model, a real system must take into account all of its requirements. There are requirements that are not definable within a prototype medium.

Exit Criteria

This step will be exited to the "clean up prototype and document" step if all rigorous components are completed. This step will exit to the "rigorous specification components" step if detail specification of any component is still needed.

3.2.8 Rigorous Specification Component

Purpose

The purpose of this step is to define all definition deliverables that require rigorous specification.

Description

All the items that cannot be documented through the model still need to be documented. The following is a list of some of the more obvious:

System Outputs. The volume, frequency, retention, and security requirements for each system output.

System Inputs. The volumes, frequency, security, and input media for each system input.

Conversion Procedure. From where and how will the new systems data base be populated?

System Logic. A complete specification of logic requirements for functions that were not or incompletely mimicked in the prototype.

System Data Base. The volume, security, integration with organization data plan, backup, and recovery specifications for the data base.

System Reliability. The performance windows, allowable downtime, and peak performance needs required by the system.

User Sites. The location of all users, distribution of allowable functions and number of users which the system will serve.

Prototyping does not exempt the necessity of cataloging these requirements. If they are not addressed now, when will they be? A production system implemented without consideration of these factors will at best prove an embarrassment and at worse be inoperative.

The prototype can be a significant aid in performing the rigorous definition process. Frequencies of inputs can be discussed with the screen on the CRT as a reference. Retention of an output report can be discussed while looking at it. Again, even the definition of rigorous components can be anchored in examples.

The rigorous components need to be entered into the same dictionary as the prototype model. This will permit a unified and coherent definition package to be produced.

Exit Criteria

This step may be exited when the components are defined.

3.2.9   Impact on Prototype

Purpose

The purpose of this step is to determine whether the additional knowledge learned from the rigorous components invalidates the model.

Description

It should be expected that proper consideration of the feasibility study and needs analysis would have provided sufficient guidance to prevent a major directional error. However, new information about a user's total needs can partially invalidate the model. There is little reason to hand off to preliminary design a model that is clearly infeasible to implement.

If the model is compromised, the revision/demonstration sequence should be reinitiated to develop a model that is satisfactory to the user in light of the additional information. To minimize the risk of this type of unpleasant surprise, detail component specification can take place in parallel with the latter iteration steps.

It should be realized that unless one is dealing with an incredibly large and complicated application, it is doubtful that the rigorous components will materially affect the model. Given the ever expanding hardware/communications technology, the user's business requirements, not physical system attributes, will be the critical path to implementation.

Exit Criteria

This step will be exited to the "prototype done" step if there is no impact on the validity of the model. This step will be exited to the "revision" step, if additional refinement of the model is required due to invalidation.

3.2.10 Cleanup Prototype and Document

Purpose

The purpose of this step is to put the model into a finished form so that it may serve as a definition baseline for the remaining steps of the system life cycle.

Description

The prototype will be of little value to anyone other than the prototypers if it is a personal undocumented system. A prototype offers two types of documentation: the vivid working model and the traditional type of documentation, that is, data element listings, record listings, and so on. Regardless of the level of automated documentation provided by the prototyping software, final documentation components like logic narratives or interprogram flow diagrams are often required.

It should be clear that most of the traditional documented components are still needed. If the prototype is to be translated to a different implementation architecture, the developer will ask the normal questions: What are the elements? What are the functions? and so on.

Some people have suggested that extensive documentation is not consistent with a prototyping approach. Though this is certainly true when one starts the process, both the prototypers and receivers of the prototype need traditional documentation to be able to refine and analyze it. We would expect that a prototyping environment with an active dictionary driven architecture will automatically deliver most of the needed documentation. A prototype without proper documentation is just another undocumented and probably nonmaintainable system.

Exit Criteria

This step may be exited when the documentation is completed.

Life Cycle Summary

The prototyping life cycle that has just been reviewed offers neither a simple solution nor an instant system. It offers a pragmatic and professional approach to developing both the user's and the developer's needs. Pragmatism requires

☐   addressing all the deliverables that are required, not just those that are directly provided by the model

☐   providing sufficient documentation to permit people other than the prototypers to understand the model

3.3   PROTOTYPING PRINCIPLES AND TACTICS

The prototype life cycle that has just been presented is challenging. Though it is, there are principles and tactics that the prototypers can exploit to minimize the challenge. If it was necessary to start each prototype as a unique undertaking, a prohibitive amount of work would be required. This, fortunately, is not the case. There are a fundamental set of principles and tactics that can tremendously leverage the entire prototyping process. By systematically applying these concepts, much of the challenge is concentrated on identifying the "exception" nature of the application. Much of the application's functionality, structure, and user interface can be predicted and often reused from other models.

Principles refer to a philosophical approach to the prototyping process. They represent a mind set on how one views the building of models. Tactics refer to specific procedural ways of conducting the prototyping process. They represent operational guidelines on how to perform the building process.

3.3.1 Prototyping Principles

The following principles should be used to enable the prototyping process.

Principle.   Most applications are derivable from a small set of system structures.

Most conventional business applications are derivatives of a few basic system structures. Most conventional systems have been solved structurally numerous times, though the immediate environment tends to camouflage the commonality. It is not necessary to start each prototype in a vacuum as an original application.

Figure 3.3 is a familiar illustration of a system flow chart. It is a convenient way to model system structure. If you review many flow charts, you will notice that a common set of structures tend to reappear.

This is an important asset to the prototyper. When commencing a prototyping project, it is not required to reinvent the system wheel, it is only necessary to identify which wheel it is. Once identified, much about the nature of that type of application is immediately known.

There are eight basic system model constructs:

Batch Edit/Update.   Batches of user inputs are grouped together on a periodic basis for input to the system.

Batch Reporting.   On a periodic basis, standard and/or ad hoc reports are derived from the data base in a batch mode.

Batch Transformation.   On a periodic basis, batch programs with extensive transform logic update the data base.

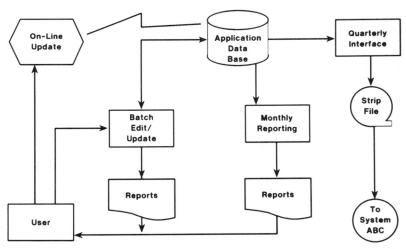

Figure 3.3 System flow charts. A review of multiple system flow charts indicates that a few basic structures repeatedly appear.

Batch Interface.   On a periodic basis, one or more batch interfaces, input and/or output, occur between systems.

On-Line Structured Update/Query.   On a periodic basis, transaction processing occurs between the user and the system.

On-Line Ad Hoc Query.   On a random basis, ad hoc requests are processed by the system.

On-Line Interface.   On a periodic basis, one or more system interfaces occur between applications in a real-time basis.

On-Line Reporting.   In response to a transaction, a report is printed either immediately or in deferred batch.

The question of interest to the prototyper is: "Which system models lend themselves to prototyping?" Figure 3.4 is a composite good candidate for prototyping. On-line transaction oriented systems with backend reporting and interfacing are good candidates. Systems with primarily batch updating and batch transformation are poor candidates. This selection criteria is basically because of the following:

Software technology has evolved to the point where building on-line models is relatively simple while batch remains quite labor intensive.

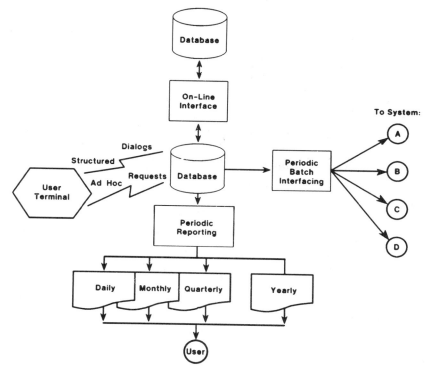

Figure 3.4   System structure conducive for prototyping. On-line transaction processing systems are ideal prototype candidates.

The benefits to be derived from prototyping lend themselves better to interactive as opposed to batch applications.

This is not particularly constraining since the thrust of most new systems is interactive. It should also be noted that the advent of distributed processing and personal computers nests and complicates the models but does not discredit them.

The points of importance for the prototyper to remember are as follows:

By recognizing the application model, an important candidacy consideration can be determined.

By recognizing the application model, natural questions and requirements come to mind.

By recognizing the application model, other systems can be identified as demonstration and component sources.

Principle. Most systems use a repetitive and well-known set of model functions.

There is a common set of basic functions that, as customized subsets, reappear in most conventional business applications. By approaching the problem with the perspective of a proven set of model functions, the problem can often be reduced:

Confirming the need for a basic function and identifying the application specific nuances.

Identifying the unique functions for the application. This can often be left for the iteration cycles when the initial model as a communications anchor is in place.

Just as system structures tend to repeat across applications, generic transform logic (especially for on-line applications), tends to reappear in many different applications. Most applications require primitive functions like:

ADD.   Add a new record to a data base.
MODIFY.   Modify an existing record in a data base.
DISPLAY.   Display an existing record in a data base.
DELETE.   Delete an existing record from a data base.

Users have common needs to look at their files with functions like:

SHOW.   Show me the keys of all *XYZ* records.
BROWSE.   Let me sequentially "walk through" a record type.
FIND.   Find me the records that meet stated conditions.

Though the phrasing of the function will be application sensitive and implementation will require considering the unique nuances of the application, a powerful base set of familiar functions does exist.

The following is a list, by no means exhaustive, of functions that are common to record management systems. Synonyms for each function are provided in parentheses.

ADD (ESTABLISH, CREATE, START, BEGIN)
MODIFY (CHANGE, ALTER, REVISE, CORRECT)
DISPLAY (RETRIEVE, REFERENCE, PROFILE, SHOW)
DELETE (REMOVE, CANCEL, TERMINATE, END)
LOCATE (FIND, GET, SEARCH)
BROWSE (TRACE)
ACTIVATE (REINSTATE)
COPY (CONVERT, MODEL, DUPLICATE)
RELATE (CONNECT, ATTACH)
STOP (HOLD).

Applications certainly have very specific functions. Functions like POST, BACKORDER, STATUS, or SUMMARIZE will have very specific meanings. Nevertheless, the availability of a widely used set of functions gives the prototyper an important edge. The user's work can be viewed in terms of familiar solutions. If not requested, the user can be prompted for the functions that experience has shown have general benefit. Lastly, incomplete models composed of basic functions can be delivered that will meet a good part of the user's needs.

The availability of model functions permits the prototype to be delivered as a shell, but a shell with highly relevant functionality. The shell can be completed during iteration as attempts to do the "exception" aspects of the application fail. The "good guess" can permit the harder definition problems to be left for the latter iterations, when many anchors will exist to help communication.

The availability of predictable functions also places the user, again, in the role of discriminating consumer. The prototyper can phrase questions in terms of "Do you need the ability to CANCEL an ORDER" or "Would you like to be able to BROWSE the OPEN ORDER FILE?" or "I assume that since we can ADD and DELETE ORDERS, we must also be allowed to MODIFY them?" If this type of questioning can be coupled with permitting the user to view other prototypes where similar functions have been implemented, the specification problem is vastly simplified.

Principle. Most input editing is derivable from a small set of edit models.

There is a well-known and repeatable set of common and generic edit criteria that is used to validate system inputs. By approaching the problem with this perspective, the problem often reduces to:

Identify the subset of generic edits required for each input.
Identify the application sensitive edits. This can often be left for the later iterations.

All data that enters an application needs to be edited for correctness. Generic edit criteria have existed for a long time. There is little point in asking a user "What are the edit rules for field *ABC?*" when one can be asking very specific questions based on experience or, even better, suggest appropriate edits.

The following is a partial list of primary generic edit practices:

1 Field tunnel editing (edits that look solely at a field).
   a Data element is a required (optional) entry.
   b Data element must be numeric.
   c Data element must be alphabetic.
   d Data element must be alphanumeric.
   e Data element may (may not) have special characters.
   f Data element must match (not match) an input mask.
   g Data element must contain (not contain) a character string.
   h Data element must be a minimum length.
   i Data element may not be a clear value.
   j Data element must be (not be) within a range.
   k Data element must be (may not be) translatable by a table.
2 Cross field editing (edits that consider multiple elements concurrently).
   a Data element must be (may not be) present if another element is (is not) present.
   b Data elements values are logically dependent on the value of other data elements.
   c Data elements value is mathematically related to other data elements.
   d Data elements value is related to a system constant such as a date.

3 Cross record editing (edit that considers an input records relationship to the data base).

    a Duplicate records, on primary key, may (may not) exist.

    b Duplicate records, on secondary keys, may (may not) exist.

    c Rules for changing fields on a record are dependent on the existence of other records.

    d Rules for adding, changing, deleting records are dependent on dates, time, cycle, or other system parameters.

Application sensitive editing can get very complicated. Extensive boolean logic and algorithms may need to be applied to specific situations. Nevertheless, the referencing of model edits permits

☐ unambiguous questioning of the user

☐ prompting of the user for edits

☐ the ability to create a working shell for the initial model with minimum specification

Principle. Application reporting is based on a four-step report model.

Generating reports from a data base can be viewed as a four-step process:

1 Select and strip data from the data base.

2 Sort each report per specification.

3 Format and edit the data for printing.

4 Print the report.

The critical factor in this sequence is the initial availability and definition of the data. Correct modeling of the data in the data base makes batch reporting via nonprocedural report generators a relatively straightforward task with extensive user participation for the later iterations of the life cycle. If a user friendly report writer and query language is provided, the user might do much of this himself.

The primary points for the prototyper are as follows:

Reporting is data base driven.

Reporting can be postponed till later iterations.

The user can directly help develop the reports.

Principle. There are a set of "correct" design structures that should be "value-added" to the prototype.

There are design practices that through experience and the test of time prove themselves to be of lasting utility to a wide range of applications. Prototyping should be value added in the sense that whether explicitly requested or not, proven design constructs, which are applicable, should be amended to the model.

One of the differences between a professional and an amateur is that through education and experience, a professional knows that certain structures are better than others. Certain structures are consistently needed and consistently deliver high utility. There are correct ways to build things.

Prototypers in many ways must be like professional architects. A client's request for a house is more in the nature of exception and special requirements: external items that the nonprofessional can relate to and is of visual importance to him. A client rarely specifies the details, the nails, the distance between supports, and so on. They assume the professional architect will of course build a correctly structured house. Similarly, users cannot be expected to explicitly state every requirement. The prototyper must value add to the user's request those parts of a system that are inherently appropriate.

Some examples of "value-added" prototyping are as follows:

Inclusion of an audit trail file.

Maintenance of control totals.

Both menu driven and command driven operational modes.

Help facility.

Standard screen headings and footings.

Date stamping of records. For more sensitive applications records may also be time-stamped, terminal stamped, and/or user stamped.

Delete processing that requires a confirmation sequence.

Consideration of ergonomics in designing screens and reports.

This list is not inclusive nor does each item need to be included in every prototype. Users need professional help in defining an application. They do not develop systems every day and need suggestions as to how a system is properly constructed.

Principle Summary

These five principles provide the foundation ideas that make rapid prototyping intellectually plausible. If every project had to be a research project, it would not be plausible to expect individuals to quickly build the unknown. Rather, these ideas provide a viewpoint that in developing a prototype we are only tailoring that which we have built many times before. This application is most likely, from the user's perspective, just another variation on a few common themes. It is perfectly plausible and reasonable to expect experienced builders to rapidly build and tailor a model of that which they have developed many times before.

3.3.2   Prototyping Tactics

The following tactics should be used to enable the rapid construction and later refinement of the prototype.

Tactic.   Data model the application by using third normal form data analysis.

There is a best way to organize, analyze, and model the applications data, third normal form (TNF) data analysis. By approaching data analysis in this way, not only is the rapid construction of a flexible data base for the model simplified, but the normalized data when coupled with the other principles, infers many of the applications requirements. In fact, one may make the following assertion: Since many applications are data driven rather than process driven, once you understand the data, you understand the needed function to drive the application. The best known technique for analyzing and modeling data is the relational TNF, which is part of Dr. Edgar F. Codd's of IBM theory on the relational data base.

TNF data analysis provides a methodical way to analyze the user's data so that:

Company Relation	
Company Code	**Company Name**
01	ABC
02	DEF
03	GHI

Location Relation		
Company Code	**Location Code**	**Location Address**
01	10	111 3rd Ave
01	12	222 6th Ave
02	13	333 8th Ave

Figure 3.5 Third normal form data analysis. Third normal form data analysis provides an optimum technique to analyze and model an applications data.

Ambiguities are resolved.

Intradata relationships are identified.

Optimum record types are identified to prevent storage anomalies.

Inter-record relationships are identified.

A logical, flexible, and process independent organization of the data is provided.

In TNF analysis, groups of data are viewed in terms of relations where a relation is a table of data (Figure 3.5). To be a normalized relation, a table must have the following attributes:

No two table rows may be identical.

The order of the rows must be immaterial.

The order of the columns must be immaterial.

Each column must have a unique identifier.

In terms of a conventional file perspective, a relation equates to a flat file, a row equates to a record, and a column equates to a field.

At the beginning of TNF analysis, the user's data is grouped as obtained from whatever sources into tables or "unnormalized" relations. TNF analysis would then consist of a four-step process:

1 Convert the unnormalized relations to first normal form (FNF) by the assignment of primary keys and removal of repeating data groups into new relations.

2 Convert FNF relations to SNF by removing partial key dependencies.

3 Convert SNF relations to TNF by resolving intercolumn dependencies.

4 Optimize TNF by combining relations that have identical keys.

The result of this process is a set of relations (tables) that can be used as an excellent first cut data base design for the prototype. In practice, before creating the data base, the relations will have to be adjusted to accommodate the physical attributes of the prototyping data base management system.

This is a very brief introduction to TNF analysis. As how to perform TNF analysis and the related issues of data base design can consume a book by themselves, the goal was just to identify the topics. For more detail on this topic, many good books exist.

If prototypes of applications with 25–30 record types with a myriad of relationships are to be built, the prototyper will need an efficient technique to help in analyzing all that data. TNF can provide a powerful technique to perform that function.

Tactic. The most productive way to build models is by component engineering.

The most efficient way to derive a needed system entity is to

☐ use an existing entity as is
☐ assemble the entity in whole from existing system entities

The ability to create, manage, and reuse components is the hallmark of any multiproduct manufacturing company. It is critical to successful prototyping that a meaningful percentage of the needs can be met by

☐ reusing pieces from previous prototypes
☐ creating new pieces with a high reusable factor for this prototype

Figure 3.6 is an example of an on-line program that has been constructed from parts. All the familiar data processing entities elements, modules, records, and so on, are first independently defined as a "part" in a storage medium (dictionary) and then assembled or plugged together

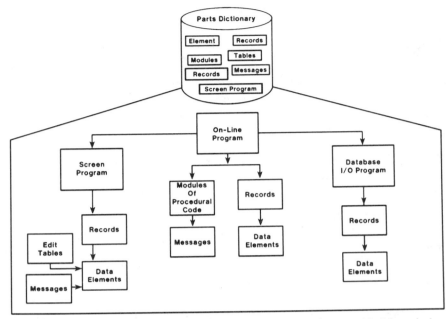

Figure 3.6 Component engineering. The on-line program is assembled from independently defined and reusable parts.

into a new part called a program. A program, then, represents the intelligent assembling of previously defined pieces. Of course, once defined these pieces can be reused as needed to form other programs. This is not possible if a program is viewed as a self-contained and local definer of itself.

To deliver "overnight" prototypes, assembly, not construction, must be the tactic. There is no time to build and rebuild the same function. The building of components not only addresses the issue of speed, but it also alleviates problems with testing, documentation, and change control.

The adoption of component engineering as a tactic is a natural result of the previous mentioned principles of system models, function models, and edit models. If, in fact, perhaps 40 specific edits could be defined, wouldn't it be highly advantageous to create small tight edit modules or tables that could be attached to input elements as required? As the requirements change, the edit pieces could be reconfigured (not rewritten) to solve the user's needs.

Building applications by the definition of parts has always been a data processing goal. Unfortunately, most third generation languages like COBOL or FORTRAN, though they have facilities to permit reusability, do not structurally force the construction of pieces. The developer must impose a parts philosophy on the software. As will be suggested in Chapter 4, when software requirements for prototyping are defined, the inherent ability of prototyping software to force the building and assembly of parts is an important selection criteria.

A parts philosophy, to be effective, should not be limited to only modules or records. Subsystems are also parts. Recurring subsystems that cross applications like

- run time help and help maintenance
- building and reporting of audit trails
- security processing

can be built in generic ways to be pluggable as needed into any prototype.

In addition to speed and quality control, components play a key role in change management. As a named entity, a part is identifiable and traceable through a prototype. When change occurs, all the places affected must be identified. Named parts can be identified for impact analysis. Programs that are self-contained solutions, do not lend themselves to change identification.

Prototyping is committed to speed, flexibility, and change. To accomplish this, leverage is required. The best possible leverage is to assemble, not to construct. Component engineering is the cornerstone of the building process for successful prototyping.

Tactic. The next best productive way to build models is by "cut and paste."

If a new component cannot be derived by assembling existing components, the next best tactic is to reuse the component base by "cutting and pasting" them into new components. This practice is commonly applied in legal offices to create customized contracts or wills from existing documents. It is more efficient to edit, rearrange, and merge proven logic structures into a new structure than to start with a blank page.

Consider the project billing system problem illustrated in Figure 1.5. Though the functions ADD was implemented 10 times in the application, it was only written from scratch once. All the ADDs used the same basic logic structures. After completing the ADD for the CUSTOMER record, it served as the basis for the PROJECT record logic although changes were needed. Since an ADD is an ADD is an ADD, the logic flow and structure could be leveraged. As the prototyping continued, a greater pool of customized functions existed from which to choose the next "cut and paste" candidate.

Cutting and pasting is applicable to most data processing entities. New records are created by deleting elements from existing records. New modules are created by selectively copying code from existing modules. New screens are initiated by copying a model screen with standard headings, footings, and formats for the project. For the project billing system, two model screens were created: one for menus and one for "doer" screens.

There are two corequisite conditions to performing "cutting and pasting." First, a component engineering philosophy must be implemented. Large intertwined pieces are not conducive to precision cutting. Secondly, a good editor who can permit the developer to concurrently juggle multiple work spaces is needed. This enables the cutting and pasting to be done conveniently and efficiently.

Tactic. Systems by Example

Communication will be optimized by permitting people to present their needs in terms of referencable anchors. The availability of a portfolio of applications exhibiting a diversity of function and presentation provides a way to establish a set of communication anchors. It is advisable to permit the user to browse the portfolio of prototypes during the analysis phase. Many conversations can then be accomplished in terms of "like screen *ABC*," "similar to report *DEF*," "same as query *GHF* except." This not only makes it easier for the user, but it also permits the prototyper the opportunity to appreciate the user's taste prior to the building of the initial model.

Data processing is one of the few businesses that does not actively show its products to the user community. Consumers regularly walk through showrooms, department stores, and supermarket aisles and gain familiarity with the types of products that are available. Only data processing seems to hesitate at showing the consumer up front the product line.

Examples are one of the best ways to communicate. The closer to actuality the example, the clearer the message. There is no reason why at the beginning of projects that users should not be demonstrated systems with similar functionality. If they are to participate effectively in the prototyping, they must be product wise.

Consider the graphic arts department in a typical company. As soon as one enters the office area, one is immediately presented with examples of the department's artwork. Examples are portrayed on the walls and display cases. The art department is making you product knowledgeable from the minute you enter the office.

Showing the user examples of existing applications is an important tactic. Once the prototype portfolio has been developed, why wait until the first cut model to get the user's imagination stimulated. Examples of all the model functions can be demonstrated. The user can learn the difference between menu driven and command driven operational modes. Function key control can be illustrated. The ergonomic power of color screens can be shown. The user can participate in the analysis discussions as a wise and informed consumer.

As is the case with any developer, an informed user is a prototyper's best friend. Though prototyping is philosophically committed to change, doing it right the first time is still preferential and cheaper. The more specific the user can be initially, the better the first guess will be. The initial prototype need not be a surprise package. It is more desirable that what is delivered in the first cut is derived for user's requests that said "like that one" or "similar to that" than only developer judgment.

Tactic. Dictionary driven software architecture

The only way to accomplish speed and flexibility is by employing a development architecture that is based on an integrated and active data dictionary. A dictionary driven architecture will not only provide for nonredundant definition of all system components but since the dictionary internally models the application, the dictionary can automatically document most of the prototype.

One of the prerequisite assumptions for making prototyping a correct strategy for performing requirements definition was the availability of highly malleable "tinker-toy" software. If you are to build high function/high reusable "tinker-toys," you must have a dictionary in which to define, maintain, and locate them.

A dictionary is a repository of the definitions of all components of a system. It stores and models the application. A special type of software dictionary is one that is both active and integrated. Integrated means that all developer tools for building system parts use the dictionary as the source for obtaining other pieces. Active means that the dictionary records all the relationships between system parts as they are built.

A software architecture based on an active and integrated data dictionary provides a complete record management system for the prototyper. All system entities and their intersystem relationships are stored in one place. High reusability and automated generation of documentation are a natural result.

Tactic. *Automated documentation*

Though a prototype does not have to be predocumented, it certainly has to be extensively documented if it is to be changeable and of value as a requirements document to the actual builders. To be of value documentation must be complete, consistent, and perfectly accurate. The only way to deliver these characteristics is mechanized documentation that uses the prototype, itself, as the documentation source.

It would be impossible for the prototyper to build concurrently the model and maintain perfect records of every relationship between all the system pieces. Yet, such documentation is critical to the iteration process. This dilemma can only be reconciled by viewing the prototype as an application data base from which complete and accurate documentation can be mechanically derived.

Tactic. *Small prototyping teams*

Prototypes cannot be built by large groups. Regardless of the size of the application, not more than three and preferably two individuals should compose a prototyping team. It is impossible to achieve speed, unity of purpose, unity of vision, good communication, and low overhead in a rapid build environment when team size exceeds three people.

As team size grows, so does the need for formal predocumentation, checkpoints, reviews, and all the other necessary control techniques to manage larger projects. Two people is an ideal size. It provides for synergy, easy communication, informal documentation, and unity of purpose. All without the need of elaborate project management overhead.

Figure 3.7   Prototyper workbench. All the software components should be easily accessible from an integrated workbench.

Tactic.   Interactive prototyper's workbench

The prototypers must be able to create the model (Figure 3.7) from an interactive and comprehensive workbench. All the software components must be executable from an interactive CRT. Such a facility provides quick feedback and continuity of activity. Both are obviously important to achieve rapid development.

Tactic.   Specification by declaration

Required software function can be derived in two ways:

Procedurally—procedural code is written that exactly states how to do the function.

Declaratively—a statement is made of what is to be done and the software decides how to do it.

Procedural specification is labor intensive, time consuming, error prone, and requires extensive checkout for correctness. Declarative specification only requires stating the requirement. If wrong, the need is restated correctly.

Declarative specification yields tremendous leverage to the prototyper. Rather than writing 20 lines of code to format a date field for printing, the prototyper declares "print as date" and all the necessary logic is automatically done. Declarative specification is highly preferable to procedural specification whenever possible.

Tactic. End-user report generation tools

Though much of the edit/update parts of the prototype would require the prototypers to construct the application itself, this is not true for the reporting segment. As reporting tools have improved in friendliness and declarative nature, end users are becoming increasingly comfortable and interested in doing it themselves.

End users who are interested and willing can build their own reports. This reduces what the prototypers have to do, increases user participation and commitment to the model, and most importantly eliminates the translation step from user to prototyper.

Tactic. Professional prototypers

To be effective, prototyping must be done quickly, be of high quality, and have a high probability of being a "good guess." This requires careful selection and development of professional prototypers. A professional prototyper should be an individual fully experienced in the entire development life cycle, fully trained and competent in the prototyping architecture, and willing to accept responsibility for delivering a high visibility product. Prototyping cannot be done casually. Casual prototypers will at best deliver passable models and at worse destroy user confidence in both the prototyping process and the credibility of the development organization.

Tactic. Developer participation in prototyping

The people who will eventually have to implement the real system from the prototype should be active participants in the prototyping process. The more they understand now, the clearer and less ambiguous the documentation they have to work with will be. This is the developer's opportunity to throw one away before building the real one. Ideally, the actual developer should participate as an active member of the prototyping team.

Tactic Summary

These 12 tactics provide specific ways to conduct the prototyping process. Rapid and accurate model building is neither an accident nor a miracle. It is the product of following specific tactics that reduce error and maximize productivity. While the previously cited principles provide a philosophical perspective on why prototyping is conceptually feasible and a viewpoint on how to look at a problem, these tactics provide the mechanisms to perform successfully the building process.

Many of these tactics are implementation architecture dependent. They are, in fact, the requirements for the prototyping software architecture. To be a valid software architecture to perform large system prototyping, these tactics must be implementable within the architecture.

Implication of Principles and Tactics

Model building is a physical process, not a logical one. By entering the prototype life cycle with these ideas, we are well prepared to handle the physical construction problem. We can leverage our knowledge and technique to its limit and concentrate during analysis on the "exception" and "special" attributes of the application. By approaching the problem in this way, there is a good chance of delivering a high quality initial model that will serve as a good anchor and, just as importantly, the prototypers will be postured to handle the potentially difficult iteration cycles.

3.4 HYBRID PROTOTYPING STRATEGIES

The prototype life cycle that was presented is a complete approach to performing requirements definition by prototyping. Some problems do not require such a formal and thorough effort. Smaller and simpler problems do not always warrant such completeness. Consequently, to achieve maximum effectiveness, prototyping needs to be varied and flexible to match the nature of each project. Many of the structured methodologies suffer criticism for being monolithic approaches that are neither sensitive nor proportional to individual problems. Prototyping should be responsive and in proportion to the immediate need.

The prototyping life cycle that has been presented implies a number of constraints on itself:

A full model will be created.

The prototypers will build the initial model.

Prototyping will commence at the definition phase.

The actual system will be built by in-house resources.

The following are some alternative approaches to prototyping that vary these constraints.

3.4.1   Prototyping of Screens Only

The presented prototype life cycle has an objective of delivering a fully working model. Working must certainly include the updating and retrieval of records from a data base. Naturally, the definition, creation, and updating of an applications file take time and effort.

An alternative approach is to constrain the prototype to mimic the system screens only. Screen programs are prepared that demonstrate the presentation and interscreen flow of each screen. Depending on the level of effort, editing and error cycling of screen inputs may also be included.

This is a perfectly logical approach to prototyping when the primary concern is the user/system interface. If the primary concern of the user is not system process logic but the external friendliness of the application, a screens only approach could be more responsive and efficient than doing a complete model.

The limits to do this approach also have to be kept in mind. Since only a part of the problem has been tested by prototyping, the remaining areas may prove to have unpleasant surprises at implementation.

3.4.2   Use of Purchased Applications as Initial Models

The presented prototype life cycle requires the construction by the prototypers of the initial model. For many applications, a supply of good models already exist. Software vendors provide purchasable systems for a wide variety of business applications. Normally, such systems are viewed as applications to be implemented into production as quickly as possible. They can just as well be viewed as rich models from which to evolve actual needs.

The success of application software, general ledger, inventory, order entry, and so on, is normally attributed to cost savings. It is hoped that the software will meet most of the user's needs and in addition be cheaper to acquire than build. The success of application software is directly attributable to its model attribute—users can examine and experiment with it before purchase.

For most large companies, vendor application software is not a solution. The vendor solution must be general and normal in its function. Most large companies are exception driven. IF THEN ELSE logic is the common construct. Consequently, though the application software may not be of immediate implementation benefit, it could be a relatively inexpensive source of proven initial models. Rather than iteration being based on an internally developed model, iteration can just as well be based on the vendor's solution. Of course the malleability of the software becomes a key selection criteria.

As productive as in-house model building may be, buying a ready-to-use model can be even more efficient. A vendor product with acceptance in the marketplace should offer at absolute minimum a set of basic functions and a shell. In certain circumstances, this could prove to be an interesting and productive alternative.

3.4.3   Prototyping During Feasibility

Communication problems and ambiguity do not commence at the definition stage. Misinformation and fuzziness are well-known attributes of the feasibility stage as well. Since prototyping relies on much of the information acquired and documented during feasibility, it can be advantageous to do limited modeling at this stage.

Modeling at this stage not only helps again to provide some communication anchors but also helps the appropriate management steering committees to approve or disapprove an application. Most companies have system review boards or system steering committees that have to decide which of the proposed systems will be built given the finite development resources. Presentations to those boards that include limited models will help them appreciate the potential benefits of each competing application.

3.4.4   Subsystem Prototyping

Problems have different attributes. Some problems are very big and will require 80–90 screens. Other problems have only a part that is high risk. Prespecification addresses the problem of how to approach a large problem by the technique of decomposition. Each process is decomposed into a logical group of subprocesses. These subprocesses can then be managed individually but in perspective.

Prototyping can certainly use the same strategy. If a problem is so large that it is too complex for a prototyping team to manage within a desired time frame, the problem can be decomposed into multiple prototypes. If only a particular subset of the application is high risk, the prototyping can be limited to only that subsystem. For another subsystem where the logic is well understood but the user is concerned with the comfort of the machine/user interface, do "screens only" prototyping.

Decomposition provides a powerful technique to break large problems into manageable units. Appropriate individual prototyping strategies can then be matched to each unit as required.

3.4.5   Prototypes and Request for Proposals

Many companies use nonstaff development resources as part of their development process. Prototypes can be of great assistance to all participants in creating a request for proposal (RFP).

A prototype, once completed, is a definition document. The target system does not have to be developed internally. An RFP to vendors for cost and time can include examination of the prototype. Additionally, existing vendor application systems can be compared to the prototype to see whether a purchase may be possible. In the case where a consultant is requested to perform a requirements definition for an application, it can be contracted that the definition should be delivered in the form of a working model. A working model will tell the user much more about the consultant's view of their need than 20 pounds of paper.

3.4.6   End-User Prototyping

There are applications for which it is perfectly feasible for an end user, with some help, to develop the prototype themselves. Most of the reason for the necessity of prototypers is that as systems get larger, record maintenance gets complex. As a consequence, reasonably complicated logic structures are necessary to check a DELETE for validity or a CHANGE for acceptability.

However, many applications consist essentially of structured and ad hoc reporting. Given the powerful reporting tools that should be part of a prototyping software architecture, it is perfectly reasonable and desirable for end users to build their own models. It eliminates all translation problems between user and implementator and places responsibility exactly where it belongs. This of course requires a user who is able and willing.

It is not meant to be overly restraining in suggesting where end-user prototyping is feasible, but a clear distinction must be made between the complexity inherent in performing certain data processing functions. The "model" for reporting is well developed and has been implemented very well by numerous vendors. Adding, changing, modifying, and deleting records in an application with 20 or more record types and 25 or more interrecord relationships is inherently complex. This remains best left to a professional builder.

Nevertheless, certain structures do lend themselves to effective end-user development. If the user is able and willing and the system structure is appropriate, letting the user perform the modeling by himself or herself can prove to be highly effective in developing a truly user satisfied system.

Hybrid Summary

These are certainly not the only other ways to approach prototyping. They do, however, serve to make the important point. Prototyping should be flexible and accommodating. It should bend to meet the needs of the situation. An unbending methodology whether prespecification or prototyping will not be used but will be viewed as a burden rather than a help.

3.5 IMPLEMENT THE PROTOTYPE

It is only within the world of data processing that the question of whether a prototype could and should be directly implemented would even be raised. Other disciplines accept prototypes for what they are—models. Only in data processing does everybody want to share the secret: "They

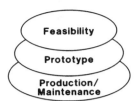

Figure 3.8   Unrealistic system life cycle. It is neither feasible nor prudent to implement a prototype directly.

say it is a prototype but we all know that it is really a system and we will be able to use it after we are done iterating." The reason for this is rooted in a perverse honesty. After delivering prototypes in the form of prespecified systems for so many years but claiming that they were not prototypes, we can now actually deliver prototypes honestly.

It is not realistic that an SDLC consisting of feasibility prototype production (Figure 3.8) will be a viable life cycle. The purpose of the prototype stage was to develop user requirements. The purpose of an actual system is to implement those requirements in the context of many other requirements. Many of these requirements or constraints are ignored or irrelevant to prototyping.

The following is a list of some of the items that are normally needed by a production application which are absent at the completion of prototyping:

- ☐ preprinted forms
- ☐ operations run books
- ☐ conversion procedures
- ☐ user documentation
- ☐ production recovery/restart procedures
- ☐ quality control review
- ☐ data base sizing
- ☐ user/machine error procedures
- ☐ test plan
- ☐ hardware/communication resources obtained and in place
- ☐ emergency backup procedures
- ☐ training procedures

Unless that which has been prototyped is trivial or a minor extension to an existing application, there is much left to be done. People have an inherent desire to want to believe in magic. Whether the technique being used is prespecification or prototyping they want it to be simple, effort-less, and painless. Prototyping does not deliver a magical solution to eliminating all the necessary steps and considerations necessary to create a system that will work in a production environment. Prototyping does not address performance issues, sizing issues, documentation issues, or training issues. Aren't they also important attributes of production systems?

Prototyping should not be initiated as a vehicle for obtaining the compressed development cycle that was referenced in Figure 3.8. Prototyping can address in an excellent manner the issues of data, functionality, and user/machine interface. It cannot, however, address many other important issues. There is no magic involved in developing professional caliber systems. If all the necessary issues are not addressed prior to conversion, they will have to be addressed after conversion at considerable more pain and cost.

Recognition of the need to address all the issues does not preclude using the prototype as the partial and incomplete basis for the target system. If the architecture required for the production system is compatible with the prototyping architecture, the decision as to whether to refine the prototype into the production system, purchase an existing vendor system, or build the production system from scratch, is an economic and technical one.

Prototyping cannot be efficiently done using any and all software. In Chapter 4, the requirements for good prototyping software will be developed. Though you may perform prototyping within one architecture because of appropriate software, the application may need to be implemented on a different target machine. In such a case, though the externals will remain constant, the insides of the prototype will have to be totally revised.

It should not be assumed that software that is appropriate for prototyping is inherently appropriate for production architectures. If a vendor was to build software specifically to support model building, requirements would include in part the following:

- [ ] completely interpretative
- [ ] high human productivity without regard to machine trade-off
- [ ] small data bases
- [ ] transaction rates inconsequential

Though these types of constraints do not hamper a prototype, they are certainly not appropriate to a production system.

The entire issue of whether you can directly implement a prototype is one of honestly admitting the realities of development. A prototype, by intent and design, is not an operational system. Attempting to implement such a production system will only result in new problems as the ignored issues surface.

3.6   A NEW SYSTEM DEVELOPMENT LIFE CYCLE

Though it has been argued that a prototype should not be directly implemented as a production system, this certainly does not mean that it cannot become the production system. Applications vary widely in the machine resources they will require in a production environment. Users vary widely in the amount of supportive documentation and training they will need. Not all systems require precision data base sizing, elaborate planning of off-site record retention, or installation of new terminal clusters. For some systems, at the completion of the prototyping cycle, they will be very close to exhibiting the necessary operational characteristics of a production version.

In situations where the prototype has evolved to a point that closely approximates the operational characteristics of the application as well as the user's view of the system, reversion to the SDLC as traditionally defined would be at best unproductive and at worse foolish. In such cases, a new life cycle has to be adopted that accounts for this reality.

Figure 3.9 illustrates a revised SDLC that acknowledges the effects of prototyping on the development process. This life cycle attempts to leverage fully that which has been completed and replaces the design sequence with an optimization/completion step to finalize bringing the prototype into harmony with any operational constraints. As prototyping and modern dictionary driven software achieve acceptance and mature, this life cycle will often become the standard approach.

Adoption of this view of the life cycle does not excuse the need to consider properly all the operational needs of a production system. However, the fact that hardware efficiencies are making many design decisions obsolete cannot be ignored. It is often much cheaper to run it and then by the 80/20 rule tune the inefficient pieces than to attempt to preoptimize the entire application.

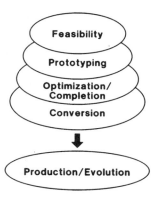

Figure 3.9 New SDLC. A new SDLC that leverages the delivered prototype is feasible.

Maintenance

If you accept the notion that a prototype can evolve into the production system, you should also be willing to entertain replacing the label of system maintenance by the notion of system evolution. Maintenance has an extremely negative connotation. It implies the minimum necessary work to keep something as is. It certainly implies that the best systems are those with zero maintenance. After conversion, they enter an idyllic state of operational bliss.

Applications that are used actively and productively by users or are part of a dynamic business environment could not possibly be maintenance free. Of course, it depends on the definition of maintenance. Much system refinement that is labeled maintenance is in reality "developmental maintenance." Since money is not allocated for development, the system is enhanced and extended under the guise of maintenance. In reality, the system is evolving in pace with the users and the business.

When you accept the ideas of prototyping, and iteration, you must realize that iteration does not magically end on implementation day. People do not stop learning or developing new insights. The business strategist and tacticians do not stop formulating new plans because the system was cut-over last week. The users and business continue to change and so will the system.

Viewing a systems growth and maturity as evolution more often honestly defines reality than calling it maintenance. If your developers are also actively engaged in "developmental maintenance," equal concern should be shown to systems with low maintenance as those with high maintenance. When we are willing to accept a new definition strategy to acknowledge the difficulties and realities of definition, we should open our minds to doing the same with maintenance.

FOUR

THE
PROTOTYPING
CENTER

There are currently three primary centers of data processing activity within a typical company (Figure 4.1):

The Application Development Center. Represents the combination of resources that are allocated to the development of new and maintenance of existing systems.

The Production Center. Represents the combination of resources that are allocated to the support and execution of production systems.

The Information Center. Represents the combination of resources that are allocated to facilitate the accessing, reporting, and analyzing of data directly by end users.

Though each center has requirements that overlap, each also has custom and specific resources to support its parochial mission. The informa-

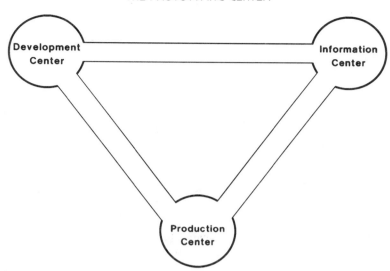

Figure 4.1   Centers of data processing activity. Most of the data processing activity in conventional business settings is encompassed by the production center, development center, and information center.

tion center makes extensive use of software products that enable end users to access data by English-like query languages, do statistical analysis, or quickly generate graphs. The production center requires software to manage the scheduling of batch runs, archiving of disk files to tapes, and securing production load libraries. Each center requires and uses center-sensitive resources.

The advent of prototyping as a formal data processing service requires the creation of a fourth data processing center, the prototyping center (Figure 4.2). Though certainly neither as large nor resource consuming as the other three centers, effective prototyping requires the allocation of a dedicated and tailored set of resources. The goals and objectives of prototyping are different from those of the other centers. Just as staff, hardware, software, and other resources are selected and groomed to facilitate the efficient operation of the existing centers, resources also need to be combined in an optimum fashion to enable the prototyping process.

In this chapter, the resources required by a prototyping center will be explained. Four topics will be addressed:

Staffing.   The selection and attributes of a prototyping staff.

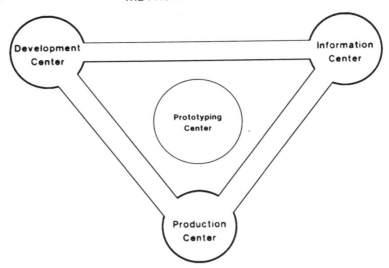

Figure 4.2 The prototyping center. The combination of resources needed to enable effective and efficient prototyping.

Hardware. The types of terminal equipment required by a prototyping staff.

Software. The selection of software that will enable the rapid building and refinement of models.

Work Atmosphere. The creation of a high productivity work environment to support the prototyping effort.

The creation of a prototyping center should not conjure up images of huge expenses or elaborate physical facilities. Much of a prototyping center is philosophical and attitudinal in nature. It is a way of conducting and organizing the business of building models.

4.1 STAFFING THE PROTOTYPING CENTER

A prototyper equates to an architect. In a world where the anticipation is that everybody will be a programmer by the use of ever higher level languages and program generators, this statement may be considered contrarious. It, nevertheless, accurately describes the role and is the correct analogy.

An architect is responsible for the total conception and unity of purpose of a new entity. The architect is carefully trained both in school and by graduated work experience in how to design and build a facility that not only meets the unique and special needs of the client, but also is inherently built and designed according to established principles.

Building models of large systems is an innovative, creative, and highly skilled activity. Building computer systems is one of the most complex activities done by people. If it wasn't inherently so complex, why would everyone be having so much trouble doing it? Determining requirements is difficult. Construction is exacting. Doing this rapidly with the goal of delivering a malleable "good guess" is not a task for the ill prepared.

The prototyper must be able to do the entire development process in miniature. The entire life cycle is collapsed into the requirements step. All the necessary tasks such as analysis, design, programming, screen generation, report generation, and human factoring must be done by the prototyper. All this must be done quickly while maintaining the application's conceptual unity and value adding to the user's requests. The prototyper, like an architect, must be a professional individual, skilled and experienced in the discipline, and be able to provide guidance to a client with imprecise needs.

Prototyping cannot be performed by large teams. The optimum size of a prototyping team is two people with perhaps a third member doing supplementary support functions for multiple projects concurrently. There are multiple reasons for the necessity of a small team:

Communication. Overhead and miscommunication must be kept to a minimum. The data processing literature verifies how communication problems grow exponentially while marginal productivity declines as team size grows. As team size passes two persons (Figure 4.3), the total number of possible interperson dialogue groups grows quickly from 1 to 22 with only a five person team. Two person teams permit personal communication and sharing of insight without the need for written documentation or extensive communication overhead.

Unity of Vision. An application should be one system, not N systems. To achieve conceptual unity and maintain its integrity in implementation, the number of creators must be kept to a minimum. A twosome will almost instinctively maintain product unity. They will constantly be integrating the next piece with the completed work.

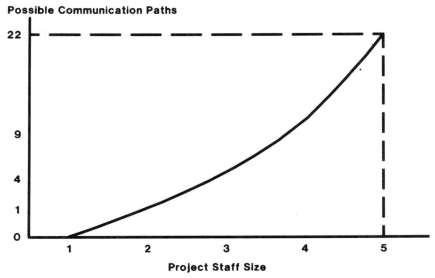

Figure 4.3 Project communication paths. As team size grows the total possible communication groups grows exponentially.

Productivity. Small teams optimize the highest productivity strategies: component engineering and "cut and paste." When you build the whole thing as you build a part, you anticipate its future uses and when you need a part, you have a mental index of the system to search. This automatic leverage is lost as team size grows.

There are, of course, compensating factors that alleviate the depth of skill required by the prototyper. Though the individual clearly must be able to do many tasks, none of them have to be performed at the expert level. The prototyper needs to be balanced and competent but not an expert in any area. Just as a family physician is versed in many disciplines of medicine at a competency level but refers the abnormal cases to the specialist, the prototyper needs to be only a general practitioner of system building and the special cases will be resolved by the area experts later in the SDLC.

The prototyper needs to design a data base that meets the user's needs, but one that ignores record volumes or response time considerations. The specialist in the data base design and communication groups will resolve those issues later. The prototyper needs to develop screens that are reasonably satisfactory to the user. The human factors consultants can refine the screens later for exactly proper spacing, use of color, wording of error messages, and so on. The difficulty of an individual having to do so many functions is alleviated by the fact that they only have to be done competently, not expertly.

The question you have of course is: Where will I acquire people with such broad skill levels? The answer is: You probably have them already. Your good system analyst and project leaders are probably doing most of the tasks already, except they're doing them within the context of traditional development. They are making design decisions on incomplete requirements daily, relating prior experience to each new situation, value-adding proven ideas to user requests, and suggesting alternate approaches to users. The same people who make the existing process work probably have the fundamental skills and attitudes to do rapid prototyping.

The people selected initially to do prototyping will most likely be incomplete in their skills. Few of us are broadly trained in everything from analysis to report debugging. There are, however, many excellent coaches in the organization. People who fundamentally understand the development process can be coached by the functional specialist in the principles and primary concepts of each discipline. The 80/20 rule probably holds again. Eighty percent of the discipline can probably be explained and learned relatively quickly. Only the last 20% takes years of learning and experience to master. For the prototyper, the 80% will probably suffice.

The basic principles of good data base design can be imparted in a few sessions. A human factors consultant can explain the basic principles of a good machine/person interface in a few hours. A quality control analyst can give guidelines on good strategies for quickly testing a program. In summary, prototypers who from conventional experience have developed a good appreciation of the development process, can be coached by the experts to achieve competence in any necessary area. Confidence and skill in execution will still require experience, but the potential staffers of a prototyping center are certainly within your organization today.

An excellent vehicle for developing an on-going prototyping staff is to apply an apprenticeship concept to the prototyping center. People become highly skilled, competent, and proficient in doing any activity by constant repetition. In conventional development, since the projects take so long or are cancelled so frequently, many analysts or developers hardly ever complete a project. Due to its rapid life cycle, a prototyping environment is an excellent milieu to train people in the complete development process. In a prototyping center, an apprentice could be exposed to perhaps four or five systems in a year. The nuances of order entry, billing, inventory, and employee record management would all be experienced from start to end. The best way to develop new prototypers is to take people with the fundamental skills and place them in an environment with continual exposure to rapid building. An apprentice working in conjunction with an experienced prototyper is an excellent way to transfer technique and knowledge.

There is no logical alternative to professional prototypers. The individuals must be able to develop rapidly a complete vision of a solution, value add to the user's requirements, and actually build it. All of this must be done at a reasonable and uniform level of competence. A division of labor approach using specialist for each function, requires too much formal communication, slows the process, and jeopardizes the delivery of a unified product.

The skill levels of a pool of prototypers only have to be proportional to the mix of problems. The experience and skills of an individual who will be addressing an 80 screen prototype need to be superior to an individual who will be responding to a request of a subsystem prototype with only six screens. Though both individuals need similar skills and experience, obviously the depth of each can be quite different. The normal distribution of the organization's problems will dictate the skill distribution needed by the prototyping staff.

Unfortunately, rapid models of significant systems cannot be built by committee. Rapidly and committee are contradictions in terms. They can, however, be built by able virtuoso performers working in an appropriate environment.

4.2   HARDWARE REQUIREMENTS

The hardware required to support a prototyping center is a function of the selected software. The software selection drives the selection of the hardware architecture. Regardless of whether the software will run on a large central mainframe or a prototyping center located minicomputer, the perspective of both the prototypers and end users to the hardware will be one of terminals.

4.2.1 Terminals

The terminal types required by the prototyping center are derived from three sources:

User Terminals. The prototyping hardware architecture must provide the same types of terminals that are conventionally used by end users.

Prototype Software Terminals. The prototyping hardware architecture must include the terminal types required to use the prototyping software most efficiently.

Print Terminals. The prototyping hardware architecture must include high speed print terminals to enable quick access to batch printouts.

Normally these terminals overlap. The same terminals used by users are the same ones used by developers. However, this does not have to be the case. Users may be normally given 24 × 80 full screen terminals. It may, however, be advantageous to give the prototypes 48 × 80 full screen terminals to facilitate using the split screen functionality of the software.

The following types of terminals would normally be expected to be used by a prototyping center:

Full Screen (Bisynchronous) Terminals. These terminals permit a full screen of data to be passed back and forth between the terminal and the computer at a time. They come in many sizes (24 × 80, 48 × 80, 24 × 132), color/no-color, number of programmable function keys, and extended hardware options (underscore, blink, bell).

Character-by-Character (Asynchronous) Terminals. These terminals transmit the data between the terminal and the computer a character at a time. They come in multiple modes (display and hard copy), different line widths (80/132 columns), and print quality.

Batch Print Terminal. This terminal serves as a high speed batch printer. In certain configurations, multiple printers may be required. One may act as a remote batch terminal to the host operating system and serve as a printing station for batch runs. Another may serve as part of a full screen terminal cluster to permit direct hard copy printing by on-line applications.

The exact selection and combination of terminals is driven by the unique combination of requirements of each environment.

4.2.2   Personal Computers

A prototyping center should consider carefully the availability of a pool of personal computers with appropriate terminal emulator software. Personal computers are reasonably portable. As a consequence of terminal emulation software, a personal computer can act as any of the other kinds of terminals and yet be portable. This provides the prototypers with two important benefits:

1   The personal computer can be taken home to permit the prototyper to work in the comfort of his or her own home without the loss of terminal function.
2   More importantly, the personal computers can be taken to user work sites to perform iterations and demonstrations. They can be left at the user's site to permit operational testing.

Personal computers may also provide the opportunity to deliver supplemental function to the prototyping effort such as word processing and graphics.

Portability does involve some problems with modems, communication speeds, and the availability of "dial up ports." All of these are solvable. Personal computers as prototyping terminals offer extended functionality, portability, and composite terminal function. They are certainly worth considering as part of any hardware architecture.

4.3   SOFTWARE REQUIREMENTS

It is obvious to anyone who has built many applications that conventional software is not very "soft." To the contrary, most software is steel-like during creation and brittle during maintenance. "Soft," not hard or brittle software, is required to enable prototyping.

A number of requirements for prototyping software have already been alluded to in the book:

- ☐ data dictionary driven
- ☐ structurally encourages component engineering
- ☐ enables "cutting and pasting" of new components from existing components
- ☐ provides an interactive prototyper workbench
- ☐ permits declarative specification rather than procedural specification
- ☐ automatically generates application documentation

These and other requirements will be analyzed in this section. First, however, the importance of integration to the prototyping process should be discussed.

4.3.1 Integration

Figure 4.4 depicts a functional view of a conventional software development architecture. The environment is nonintegrated. Regardless of the individual merit and capability of each component:

Bridges must be built between components to permit communication.
Redundant effort is required to redefine common entities to each component in its unique language.
There is no method to insure consistency between components.
There is no way to identify the usage of an entity across components.
The entire process of building intercomponent bridges and redefining common entities is extremely labor intensive, time consuming, and error prone.

It is clear that it is impossible to perform rapid prototyping in such an environment. Even if by heroic effort an initial model would be created, the attributes of a nonintegrated environment, labor intensiveness, redefinition, and error proneness, would prohibit rapid refinement. Systems that are to be built in an architecture where each component is deaf, mute, and blind to its associated components mandates careful prespecification.
Figure 4.5 depicts an integrated software development architecture.

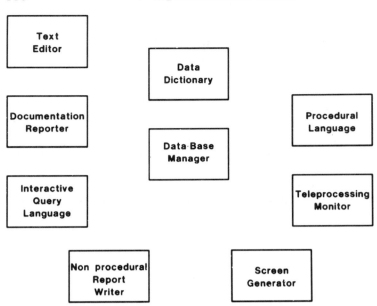

Figure 4.4 Nonintegrated software environment. Conventional software architectures are labor intensive and error prone.

Though the same generic components are required, they are integrated through a single definition point, a data dictionary. The definitions of all components are entered into a single dictionary. As a by-product of the building process, creating a program out of modules, records, messages, and screens, the dictionary is automatically updated with all the relationships established between them. Once defined to the dictionary, any entity by declaration is known to any and all other software components. Since the dictionary is active, all the relationships between entities are dynamically and perfectly maintained.

An integrated and active data dictionary provides a record management system for the prototyper. In this case, the records being maintained are those of elements, records, screens, and so on. The records are updated from two sources:

The prototyper directly defines new entities interactively. When a definition involves relating multiple entities, relationship records are established in the dictionary to record the relationship.

The integrated software components update the dictionary as part of the building process.

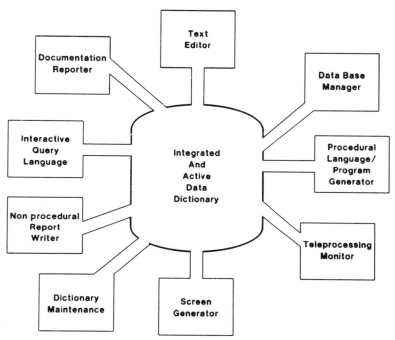

Figure 4.5 Integrated software architecture. All the software components communicate to each other through an active data dictionary.

An active and integrated dictionary negates the problems of a noninte-grated environment. Communication between components is done by a single definition point. Redefinition is never required and the associated opportunities for inconsistency errors are eliminated.

It is only possible to do professional prototyping within the context of an integrated software architecture. Only within such an architecture is the speed, flexibility, automation, documentation, and accuracy required achieved. Integration is the first and foremost software requirement of prototyping.

Prototyping, as a consequence of this requirement, has only recently become practical or possible. Prior to the last few years, no such software existed. Vendors supplied excellent components, but only components. Even when a vendor supplied multiple pieces of the full software architecture, they were developed as islands. As a result, software building was extremely labor intensive since integration was a completely manual and labor intensive effort.

Such is no longer the case. Many products are being introduced by software vendors that are integrated development systems. A single control and definition point, the dictionary, actively manages all the components. Integration has completely revised the human/machine effort ratio and made effective prototyping a practical reality for the first time.

4.3.2 Software Requirements Definition

The purpose of this section is to create a schedule of desired functionality for prototyping software. To my knowledge, no vendor has actually created a software product with the intent that its primary usage be the building of models as opposed to production applications. As prototyping achieves widespread industry acceptance, this will probably change. Specific products will be developed with the expressed goal of facilitating the building of application models. In the interim, the issue is one of intelligent selection of the available products.

Most conventional software products emphasize their production center efficiencies. They emphasize their ability to minimize resource utilization and provide run time efficiency under high volume conditions. For prototyping software, production application performance characteristics are of secondary importance. Most prototypes will manage small data bases in terms of record volumes and a large terminal network might be two terminals. The software's strength to optimize the productivity of the prototyper is the prime consideration. The tradeoff decisions made by the software creators should consistently trade human effort for machine effort. It is only a model that is being built. It does not have to run optimally. The key mental questions to keep asking oneself when looking for prototyping software are: Does this feature enable the prototyper's productivity or does it hinder it? Could this function be performed mechanically by more intelligence in the software?

The following features would be highly desirable of software that was to be used to build application models.

1 *All the software components should be integrated by an active data dictionary.*

The prototyping software should be based on an integrated data dictionary architecture. All entity definitions and relationships must be main-

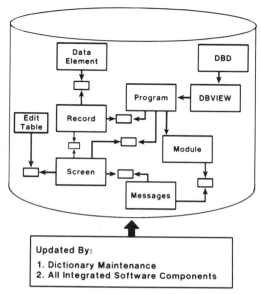

Figure 4.6 Integrated data dictionary. The dictionary, in addition to serving as the source of entity definition, "models" the structure of the application.

tained in the active dictionary. Both the prototyper and the component software products should access the dictionary so that it maintains a current "model" of the system. Figure 4.6 illustrates these concepts.

The following system entities are typical and are defined as follows:

Element.   Defines a primitive data element.

Record.   A collection of data elements.

Module.   A collection of procedural code.

Message.   A narrative to be sent to a user sitting at a screen.

Edit Rule Table.   A set of edit criteria to be applied to validate an element.

Data Base.   A definition of a data base with the records and associated inter-record relationships.

Data Base View.   A subview of a data base.

Map.   A definition of a "3270" type terminal screen.

Program.   An executable program.

This dictionary entity structure becomes the record management system for the prototypers. Whether they directly update it by the dictionary maintenance component or the architecture components (program generator, screen generator, etc.) update it as a by-product of their execution, it is required that the dictionary maintain a current and accurate representation of the system.

2   *The prototyping software should be complete and self-contained.*

All the components that are necessary to generate a complete application should be encompassed by the prototyping architecture (Figure 4.5). The software should contain the following types of components:

- ☐   an active and integrated dictionary
- ☐   a high function data base management system
- ☐   a very high level procedural language for generating both batch and on-line programs
- ☐   a teleprocessing monitor
- ☐   an ad hoc query language for both batch and on-line queries
- ☐   a nonprocedural report generator
- ☐   a documentation generator
- ☐   a text editor

This requirement is incestuous with the prior requirement of an integrated architecture. If the prototyping software is to be able to generate complete models, it must be complete in the components it provides the prototypers. If it is to be integrated, all components must come under the control of the dictionary.

3   *The software architecture should be both menu driven and command driven.*

Prototypers of different skill levels and product familiarity will have to use a wide range of software components. Menu driven dialogues are highly advantageous in helping a novice or when doing an infrequent activity. Conversely, once a high skill level is achieved, walking through a set of menus is both tedious and unproductive. Command driven control of the software component is then advantageous. Both menu driven and command driven software provides a flexible match to the immediate skill level of the prototyper.

4   *The prototyping software by its inherent structure should accommodate component engineering.*

The implementation of an architecture that supports component engineering is accomplished by making a clear distinction between raw entities and assembled entities. Most third generation languages make no such distinction. Elements can be externally defined in a copy library or they can be locally defined within a program.

A prototyping architecture should make such a clear distinction. A program should be viewed as an assembled entity (refer to Figure 3.5). All the associated entities that make up the program should be independently defined prior to program definition. Program definition then becomes one of declaring what set of dictionary entities are to be assembled. Of course, since these entities exist independently, they can be reused as needed by any other programs.

5   *The prototyping software by its inherent structure should support a "cut and paste" approach to building new parts.*

This requirement is a natural corollary to the component engineering requirement. It is much faster and less error prone to build new pieces from existing pieces than to start from nothing. To support a "cut and paste" philosophy, the dictionary must provide multiple indexes to find pieces (the dictionary must be able to serve as a parts directory) and a powerful text editor to manipulate and manage editing multiple pieces concurrently. This would require that if the editor component of the software was not directly part of the dictionary, it would still have interactive access to the dictionary.

6   *The prototyping software by its inherent structure should support specification by declaration.*

Specification by declaration is the act of stating what you want done but leaving it up to the software to determine how to do it. This type of

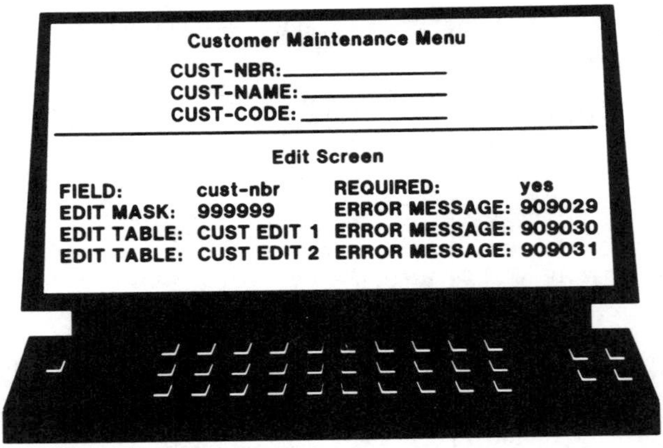

Figure 4.7   Specification by declaration. It is more productive and less error prone to declare required function than to procedurally code required function.

specification off loads a great deal of procedural code writing to the prototyping software.

Figure 4.7 illustrates this principle. The top part of the figure represents a CRT screen that has just been painted by the screen generator component. The lower half of the figure shows an "edit screen" where detail is being requested about a particular field "CUST-NBR." Lower case letters indicate prototyper inputted data.

Needed function such as

- ☐   field mask edit
- ☐   field validation against edit rules
- ☐   presentation of error messages

will all be done automatically by the screen program at execution based on the declaration. This obviously will avoid coding a tremendous amount of tedious, redundant, and error prone procedural code.

7   *The prototyping software should provide a prototyper workbench environment.*

The prototyper should be able to sit at his or her work station, a full screen terminal, and conveniently be able to control each software component interactively (and batch) and move between components as required. Function key assignments, syntax, product logic rules, and so on, should be as consistent between products as possible. Just as the dictionary unifies and integrates the software internally, the prototyper, as user, should have a unified and integrated interface to the entire architecture.

8 *The prototyping software should provide end-user facilities.*

It would be highly advantageous to the prototyping effort if the end users could perform some of the prototyping. Not only does this accomplish accruing their active participation, but it also completely eliminates any translation problems between user and implementor. The query and report writer facilities should be graduated in usage ability so that an end user could use them to generate reports, batch and on-line.

9 *The prototyping software should be able to automatically document the application.*

Whether the prototype will be used only as a definition document or will become the actual system, it is necessary that it be documented. In fact, it is necessary that to avoid translation problems it must be documented perfectly.

This requirement should be deliverable since the entire prototype is modeled in the data dictionary. All the attributes of each entity and their interentity relationships are defined in one place. The data dictionary has a complete set of current and accurate records on the prototype. Since a data dictionary is itself just another data base, it is subject to the same reporting mechanisms as any other data base.

Mechanically, one would expect a documentation generator (dictionary reporter) to be able to extract the following types of documentation reports from the dictionary:

Documentation Summary.   An alphabetical listing by entity type containing every application entity and a short description.

Data Element Glossary.   An alphabetical listing of every data element used by the prototype. All element attributes, such as description, picture, clear value and synonym names, and interentity relationships, such as records used in, should be contained in the listing.

Record Glossary.   An alphabetical listing of every record used by the prototype. All record attributes, such as description and record length, and interentity relationships such as element composition, where used in programs, screens, data base views, and data bases, should be contained in the listing.

Edit-Rule Glossary.   An alphabetical listing of every edit rule used by the prototype. All edit-rule attributes, such as description, type (range table, translate table, mask table), and rules details, should be contained in the listing.

Module Glossary.   An alphabetical listing of every module used by the prototype. All module attributes, such as description, language types, source code, and interentity relationships, such as programs where used, should be contained in the listing.

Message Glossary.   An alphabetical listing of every message used by the prototype. All message attributes, such as message narrative and description, should be contained in the listing.

Program Glossary.   An alphabetical listing of every program used by the prototype. All program attributes and interentity relationships such as records used, screen used, and data base view used, should be contained in the listing.

Data Base View Glossary.   An alphabetical listing of every data base view used by the prototype. All data base view attributes, such as description and input/output permission level (read only or update), and interentity relationships, such as programs where used and owner data base, should be contained in the listing.

Data Base Glossary.   A listing of the data base definition with all the associated record types and records relationship rules.

Map/Element Cross Reference.   An alphabetical listing by screen of all elements which appear in that screen.

Element/Map Cross Reference.   An alphabetical listing by data element of all screens in which the element appears.

Module/Element Cross Reference. An alphabetical listing by module of all data elements that are referenced in that module.

Element/Module Cross Reference. An alphabetical listing by data element of all modules in which the data element appears.

Program/Program Sender Cross Reference. An alphabetical listing by program name of all the other programs to which this program passes control.

Program/Program Receiver Cross Reference. An alphabetical listing by program of all the other programs from which this program receives control.

Module/Message Cross Reference. An alphabetical listing by module of all messages issued by the module.

Message/Module Cross Reference. An alphabetical listing by message of all modules that issue it.

Element/Edit-Rule Cross Reference. An alphabetical listing by data element of all the edit-rule tables that are used to edit it.

Edit-Rule/Element Cross Reference. An alphabetical listing by edit-rule table of all the data elements that use it.

Element/Message Cross Reference. An alphabetical listing by data element of all the messages associated with the element.

Message/Element Cross Reference. An alphabetical listing by message of all data elements that use it.

Screen Definition Report. A complete hard copy definition of each screen with the screen image, and technical attributes of each screen's literals and variables.

Report Definition Report. A facsimile image of each hard copy report.

Entity Orphan Report. An alphabetical listing by entity type of entities that are defined but not used by the prototype.

ENTITY TYPE   ENTITY NAME   SUMMARY DESCRIPTION

PROGRAM	PETED057	REFERENCE CUSTOMER
PROGRAM	PETED058	DIALOG FOR PROJECT REC SHOW
PROGRAM	PETED059	ADD BILL HEADER
PROGRAM	PETED060	ADD BILL DETAIL
PROGRAM	PETED061	DISP BILL
PROGRAM	PETED062	REFERENCE BILL DETAIL
PROGRAM	PETED063	REFERENCE PARTICIPATION
PROGRAM	PETED064	CANCEL BILL
PROGRAM	PETED065	REINSTAT BILL
PROGRAM	PETED066	REFERENCE PACREF
PROGRAM	PETED067	DIALOG FOR SHOW CUSTOMERS

Figure 4.8 Documentation summary. An alphabetical listing of each prototype entity.

Entity Changed Report. An alphabetical listing by entity type of all entities that have been added or changed since a specified date.

Figures 4.8–4.11 give examples of the following reports:

Figure 4.8 exemplifies the documentation summary of the program entity.

Figure 4.9 exemplifies a data element glossary report.

Figure 4.10 exemplifies a message to data element cross reference.

Figure 4.11 exemplifies a screen definition report.

All of these 26 reports should be mechanically generatable from the dictionary and as many as possible should be available interactively by querying the dictionary. The goal of all these reports is to document completely the prototype. This documentation is as much required by the prototyper as by the eventual recipients of the model. Without this type of documentation, change management would be a hit and miss proposition. With this type of documentation, change is a methodical and disciplined process. The impact of changing any entity in the entire prototype can be perfectly assessed by analyzing the appropriate reports.

A complete documentation package that would be handed off at the completion of prototyping would require additional documentation that cannot be generated mechanically. Manually created documentation might include

- ☐ data base design diagram
- ☐ program structure diagrams
- ☐ system flow charts

Prototyping often conjures up a vision of rapid application building without the overhead of documentation. Certainly manually prepared documentation is overhead, expensive, and always inaccurate. Mechanized documentation from an integrated dictionary is always complete, consistent, and correct. Before attempting to do prototyping of applications without the types of mechanized documentation described here, you should seriously try to answer two questions:

1   How will the prototypers be able to analyze the prototype during iteration for the impact of change?
2   How will anybody at the completion of the prototyping be able to analyze it for creating the real system?

The simple reality is that a prototype, as it grows in size, takes on the same complexity attributes of any real system. This requirement for mechanical and complete document is extremely important.

10   *The prototyping software must support the normal terminals used by the end-user population.*

This requirement is self-evident. If all the users are only given asynchronous terminals, prototyping software, no matter how good, is irrelevant if it has facilities for only managing full screen terminals. Similarly, if users are evolving to full screen colored terminals, prototyping software that doesn't permit manipulation of the color attribute flags is of qualified value. The point is obvious. The prototyping software must be in synchronization with the user populations anticipated hardware environment.

11   *The data base management component of the prototyping software must be able to model all common file structures.*

DATA DICTIONARY REPORTER C8207M
ELEMENT REPORT

```
************************************************************************************
                                                            BUILD ---- DATE --------
ELEMENT NAME                                                CODE UPDATED CREATED
************************************************************************************
```

ELEMENT NAME	BILLENT-NAME				
RECORD NAME	BILLENT	VER	IDD BUILT		
RECORD NAME	BILLENT	VER	IN SCHEMA PETESCHM VER	1	
RECORD NAME	PETE7-CA	VER	IDD BUILT		
RECORD NAME	STD-SCREEN	VER	IDD BUILT		
RECORD NAME	BILLENT-W1	VER	IDD BUILT		
RECORD NAME	SEL-BILLENT	VER	IDD BUILT		

BILLENT1-CODE	VER 1			D	11/24/82
PREPARED BY	AT12690A				
DESCRIPTION	CODE DENOTES UNIQUE BILLING ENTITY		CODE DENOTES UNIQUE BILLING ENTITY		
SYSID	PETE				
PETE-MESSAGE	DC900157				
PETE-TABLE	PETEE001		ZERO IS INVALID		
VALUE ZERO		DISP			
WITHIN GROUP	BDP-KEY	VER		PRIMARY GROUP	
WITHIN GROUP	FROM-CUST-KEY	VER		PRIMARY GROUP	
SAME AS	BILLENT2-CODE	VER			
PICTURE	9	DISPLAY	1		
ELEMENT NAME	BILLENT1-CODE				
RECORD NAME	PACREF	VER	IDD BUILT		

RECORD NAME	PACREF	VER	IN SCHEMA PETESCHM VER	1	
RECORD NAME	CUSTREL	VER	IDD BUILT		
RECORD NAME	CUSTREL	VER	IN SCHEMA PETESCHM VER	1	
RECORD NAME	PETE1-CA	VER	IDD BUILT		
RECORD NAME	PACREF-W1	VER	IDD BUILT		
BILLENT2-CODE	VER 1			D	11/24/82
PREPARED BY	AT12690A				
DESCRIPTION	CODE DENOTES UNIQUE BILLING ENTITY		CODE DENOTES UNIQUE BILLING ENTITY		
SYSID	PETE				
PETE-MESSAGE	DC900157		ZERO IS INVALID		
PETE-TABLE	PETEE001	DISP			
VALUE ZERO					
WITHIN GROUP	PROJ-BASIS-YR-KEY	VER		PRIMARY GROUP	
WITHIN GROUP	FROM-ACCT-KEY	VER		PRIMARY GROUP	
WITHIN GROUP	PROJ-GRP	VER		PRIMARY GROUP	
SAME AS	BILLENT-CODE	VER			
PICTURE	9	DISPLAY	1		
ELEMENT NAME	BILLENT2-CODE				
RECORD NAME	BASIS	VER	IDD BUILT		
RECORD NAME	BASIS	VER	IN SCHEMA PETESCHM VER	1	
RECORD NAME	BASYR	VER	IDD BUILT		
RECORD NAME	BASYR	VER	IN SCHEMA PETESCHM VER	1	
RECORD NAME	ADMIN	VER	IDD BUILT		
RECORD NAME	ADMIN	VER	IN SCHEMA PETESCHM VER	1	
RECORD NAME	LOAD	VER	IDD BUILT		
RECORD NAME	LOAD	VER	IN SCHEMA PETESCHM VER	1	

Figure 4.9 Data element glossary report. A detailed definition of each data element used in the prototype.

129

DATA DICTIONARY REPORTER C8207M
ATTRIBUTE REPORT

****

| CLASS/ATTRIBUTE | | --------DATE-------- | | ATTRI DELETION | | |
| | | UPDATED | CREATED | A S | LOCK | |

****

ELEMENT	CHANG-YEAR			VER	1
ELEMENT	WORK1-YEAR			VER	1
ELEMENT	WORK2-YEAR			VER	1
ELEMENT	WORK3-YEAR			VER	1
ELEMENT	WORK4-YEAR			VER	1
ELEMENT	WORK5-YEAR			VER	1
ELEMENT	TODAY-YEAR			VER	1

PETE-MESSAGE   A

DC900011   11/12/82   OFF

PREPARED BY					
ELEMENT	AT12690A			VER	1
ELEMENT	MODEL-YEAR			VER	1
ELEMENT	ACCT-YEAR			VER	1
ELEMENT	CNTL-YEAR			VER	1
ELEMENT	PCREAT-YEAR			VER	1
ELEMENT	PCHANG-YEAR			VER	1
ELEMENT	BILL-YEAR			VER	1
ELEMENT	PTODAY-YEAR			VER	1
ELEMENT	YYMMDD-YEAR			VER	1

EDIT-MESS

DC900012		08/04/82	07/28/82	A	OFF
PREPARED BY	AT12690A				
REVISED BY	AT12690A				
ELEMENT	BIRTH-DATE	VER	1		
ELEMENT	EFF-CHNG-DATE	VER	1		
ELEMENT	SERVICE-DATE	VER	1		
ELEMENT	XXXX-DATE	VER	1		
ELEMENT	XXXX-R-DATE	VER	1		
ELEMENT	XXX-DATE-GRP	VER	1		
ELEMENT	BLOOD-TEST-DATE	VER	1		
ELEMENT	BLOOD-TEST-R-DATE	VER	1		
ELEMENT	BLOOD-TEST-DATE-GRP	VER	1		
ELEMENT	PART-BIRTH-DATE	VER	1		
ELEMENT	PART-BIRTH-R-DATE	VER	1		
ELEMENT	PART-BIRTH-DATE-GRP	VER	1		
ELEMENT	STRESS-TEST-DATE	VER	1		
ELEMENT	STRESS-TEST-R-DATE	VER	1		
ELEMENT	STRESS-TEST-DATE-GRP	VER	1		
ELEMENT	SKIN-FOLD-DATE	VER	1		
ELEMENT	SKIN-FOLD-R-DATE	VER	1		
ELEMENT	SKIN-FOLD-DATE-GRP	VER	1		
ELEMENT	PROGRAM-DATE	VER	1		
ELEMENT	PROGRAM-R-DATE	VER	1		
ELEMENT	PROGRAM-DATE-GRP	VER	1		
ELEMENT	VISIT-DATE	VER	1		

Figure 4.10 Message to data element cross reference. A cross referencing listing itemizing all data elements that share an error message.

The file structures used for most conventional business applications can be categorized as keyed (inverted), hierarchical, network, or relational. Since the prototyping center cannot predict what problems it will receive, the data base management component must be able to simulate all the structures. This is not particularly difficult since depending on the amount of procedural programming you are willing to do, almost any structure can mimic the others.

The issue is really ease of definition, ease of change, and function:

Ease of Definition. This encompasses the steps and complexity of defining the data base structure.

Ease of Change. This encompasses the steps and complexity of changing the data base structure.

Function. This encompasses the amount of rules that can be specified to the data base manager to off-load procedural programming from the prototyper to the data base software.

Unfortunately, these requirements are often not complementary. Ease of definition and change usually result in lack of functionality. Powerful functionality usually requires more complex definition and restructuring steps. If procedural coding is to be off loaded to the data base manager, rules must be declared to it.

To evaluate the trade-off, one must consider what the typical changes made during prototyping might be that effect the data base:

☐   addition of new fields
☐   revision of a fields definition (i.e., change in size)
☐   removal of an existing field
☐   addition of a new record
☐   removal of an existing record
☐   establishment of new relationships between records
☐   removal of relationships between records

How you would do these activities during iteration should be carefully understood. You want to be able to do it fast, accurately, and minimize the rippling on the baseline prototype.

REPORT FOR MAP   PETEM001

DEVICES: 24X80, 32X80, 43X80, 27X132

```
      5   10   15   20   25   30   35   40   45   50   55   60   65   70   75   80   85   90   95  100  105  110  115  120  125  130
**********************************************************************************************************************************

                              SPROJECT BILLING SYSTEM                                    SPETEM001

                              SFUNCTION: S .......-(CURSOR)........              S .........
                                         S ............

            SSELECT      SBILLENT     SBILLENT
                         SCODE        SNAME

            S.S          S.           S ............
            S.S          S.           S ............
            S.S          S.           S ............
            S.S          S.           S ............
            S.S          S.           S ............
            S.S          S.           S ............
            S.S          S.           S ............
            S.S          S.           S ............
            S.S          S.           S ............


SENTER - EXECUTE FUNCTION                   *** PF KEYS ***
S............S...........                      S..........S
S............S...........                      S..........S
S............S...........                      S..........S
           S ..........                                        S ..........

**********************************************************************************************************************************
      5   10   15   20   25   30   35   40   45   50   55   60   65   70   75   80   85   90   95  100  105  110  115  120  125  130
```

(continued)

REPORT FOR MAP   PETEM001

DEVICES: 24X80,   32X80,   43X80,   27X132
USING RECORDS:
    STD-SCREEN   VERSION   1
    SEL-BILLENT   VERSION   1
    PETE7-CA   VERSION   1
WCC: NOALARM, UNLOCK KEYBOARD, RESET MODIFIED, NOPRT, NLCR
PANEL PETEM001-OLMPANEL   VERSION   1

PFLD: OLMPD-0027   AT (1,28)
       ATTRIBUTES = (NUMERIC,PROTECTED,NONDETECTABLE,DISPLAY,NOMDT,
                     NOBLINK,NORMAL-VIDEO,NOUNDERSCORE,
                     NOCOLOR)
       NODELIMIT
       LITERAL STRING

PFLD: OLMPF-0004   AT (2,72)
       ATTRIBUTES = (NUMERIC,PROTECTED,NONDETECTABLE,DISPLAY,NOMDT,
                     NOBLINK,NORMAL-VIDEO,NOUNDERSCORE,
                     NOCOLOR)
       NODELIMIT
       LITERAL STRING

PFLD: OLMPF-0005   AT (3,30)
       ATTRIBUTES = (NUMERIC,PROTECTED,NONDETECTABLE,DISPLAY,NOMDT,
                     NOBLINK,NORMAL-VIDEO,NOUNDERSCORE,
                     NOCOLOR)
       NODELIMIT
       LITERAL STRING

```
PFLD:  OLMPF-0006                          AT ( 3,41)
       ATTRIBUTES = (NUMERIC,PROTECTED,NONDETECTABLE,DISPLAY,NOMDT,
                     NOBLINK,NORMAL-VIDEO,NOUNDERSCORE,
                     NOCOLOR)
       NODELIMIT
DFLD:  P-FUNCTION-NAME                                      OF STD-SCREEN
       OPTIONAL
       INPUT:    JUSTIFY LEFT, DATA YES, PAD NO
       OUTPUT:  BACKSCAN NO, DATA YES

PFLD:  OLMPF-0007                          AT ( 3,72)
       ATTRIBUTES = (NUMERIC,PROTECTED,NONDETECTABLE,DISPLAY,NOMDT,
                     NOBLINK,NORMAL-VIDEO,NOUNDERSCORE,
                     NOCOLOR)
       NODELIMIT
DFLD:  P-DIALOG-NAME                                        OF STD-SCREEN
       OPTIONAL
       INPUT:    JUSTIFY LEFT, DATA YES, PAD NO
       OUTPUT:  BACKSCAN NO, DATA YES

PFLD:  OLMPF-0008                          AT ( 4,41)
       ATTRIBUTES = (NUMERIC,PROTECTED,NONDETECTABLE,DISPLAY,NOMDT,
                     NOBLINK,NORMAL-VIDEO,NOUNDERSCORE,
                     NOCOLOR)
       NODELIMIT
```

Figure 4.11 Screen definition report. A complete technical specification of a screen layout.

135

An analysis of available data base managers would indicate that none of them is perfect. The practical problem is to select the best candidate given the limitations of each.

Network Model

By definition, applications for which prototyping offers the greatest return on investment are structured problems. Network data bases support structured problems very well and have the practical benefit of being a likely target architecture. A potential problem with a network model is the need to restructure the data base if requirements change. Consequently, only a network data base manager with superior restructure utilities should be considered.

Hierarchical Model

Hierarchical data base managers are obviously inadequate. Many problems are network in nature and imposing network structures on hierarchical data bases is extremely complicated.

Keyed (Inverted) Model

Inverted data base managers tend to be very flexible at adding and changing data base structure but usually do not permit rule specification. They also have difficulty in handling a number of practical situations encountered by "structured" applications:

1. They do not easily support cyclic (bill of material structures) where there is a many-to-many relationship between two records of the same type.

2. They do not easily handle situations where the relationships between records is nonkey related, that is, last-in-first-out (LIFO) or first-in-first-out (FIFO) relationships. These types of situations require the creation of artificial keys with associated procedural logic to maintain them.

3. They do not easily support "backward browsing." Since indexes are usually forward pointers, to browse backward through a record type involves creating a complement to the index field, declaring an additional key on this artificial field, and then to browse backward you must procedurally browse forward on the artificial field that is the complement of the actual key (understand?).

Relational Model

Relational data base managers are a popular topic. It is not clear that a true relational data base manager exists. Even if it did, however, it is not obvious how appropriate it would be for prototyping "structured" applications:

1. Relational is actually a theory for defining data as a process independent resource. Data is modeled alone for its own sake without regard to process requirements. Though this is certainly a powerful way to model strategically a corporation's data as a resource, prototyping deals at a much more pragmatic level.

2. By definition, relational data bases have only implicit relationships between record types. Though this is excellent for revising the definition of records and their interrelationships, it imposes the responsibility for maintaining all record relationships procedurally on the programmer. Though in the context of a corporate approach to data administration this might be strategically correct; for prototyping this is questionable.

3. Relational data bases cannot support cyclic (bill of materials) type structures. Such a structure requires a JOIN between two records of the same type. By definition, no two columns in a relation can be the same field. Cyclic structures are surprisingly common in structured applications.

4. Most of the literature on relational data base management systems emphasize the powerful retrieval capability of the data base manipulation language. Though the SELECT, PROJECT, and JOIN retrieval operators are indeed very powerful, they do little to aid in insuring inter-record integrity during maintenance (ADD, CHANGE, DELETE, and so on) activity. For structured applications, the problems of maintaining the records and the integrity of the inter-record relationships is often much more difficult and challenging than retrieving the data. Unless the data base is maintained, there will be very little for the powerful relational operators to retrieve. As currently available, relational data base managers appear most suitable for applications that accentuate retrieval, analysis, and inquiry applications. These are not the functions that often pose the most difficulty to structured problems.

This analysis emphasizes issues of concern with each data base manager. The problems with each can be made clearer by considering a "nontoy"

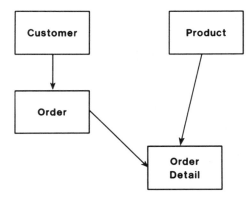

Figure 4.12 Toy data base problem. Straightforward problems can be solved well by all data base managers.

problem for which prototyping would be an ideal approach. Figure 4.12 is a typical "toy" problem that could be solved quite well by any candidate data base manager. It is straightforward, has few record types and relationships, and diagrams very nicely.

Figure 4.13 is not a "toy" problem. It's extremely complicated, has 30 or more record types, 40 or more relationships, and does not fit so nicely on the page. It, however, clearly needs a prototyping approach to requirements definition. Before investing a huge sum of money in building such an application, we would absolutely want to minimize risk and uncertainty.

The data base portrayed in Figure 4.13 is a statistics record management data base for a sports league. Specifically it maintains the season records for a professional ice hockey league. All the required data about teams, leagues, players, games, and seasons is maintained in one integrated data base. Such a data base provide multiple functions to the league and teams:

All the statistics and data about teams, players, and so on, is accumulated in one place.

The data base can serve as the source for the generation of data for league year books, sports encyclopedias, and team year books.

The data base can serve as the source for performing postgame analysis by coaching staffs.

The data base can serve as the source to support a play-by-play color commentator on television or radio.

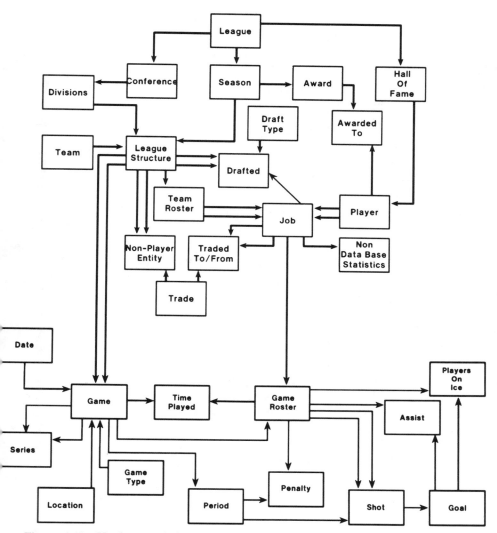

Figure 4.13 Hockey statistics data base. Nontoy data base problems like this require a powerful data base manager to structure properly.

Though a generic problem common to any sports league, the data base required to support it is surprisingly large and complex:

Cyclic bill of material structures are common.

The inter-record verification required to support maintenance activities is extensive.

Many of the inter-record relationships are nonkey oriented. Rather, they are either relative or LIFO.

The question that needs to be answered is as follows: Which data base type will be most appropriate to prototype problems like the hockey statistics data base? Consider the potential difficulties with each of the candidates.

Hierarchical.   Clearly a solution with no hope of success. Properly modeling the data base and programmatically maintaining the record JOB would be an almost impossible task.

Relational.   A relational data base manager would have severe difficulty in modeling the eight cyclic structures. In addition, since all relationships are implicit, the relational data base manager would be of little service in assuring inter-record integrity during maintenance operations. A procedural programming mistake could leave a GOAL orphaned without the GAME or GAME ROSTER that must coexist with it.

Inverted.   An inverted data base manager would require the concatenation and propagation of related record keys throughout the data base. This would require exacting programming and be a maintenance nightmare if a key changed. Many of the records like PENALTY, SHOT, and GOAL are involved in LIFO relationships. Though creating artificial keys to uniquely identify each one may be feasible at ADD time, it would require extensive and complicated procedural logic to maintain relative key integrity under corrective maintenance.

Network.   Though there would be no initial problem in designing the data base, restructuring would be a potential difficulty as requirements evolve.

The point is not to be negative toward relational, inverted, hierarchical, or network data base managers (they are the only choices) but to clearly distinguish the difficulties with each when posed with a significant problem. The hockey data base is the type of problem for which prototyping provides the greatest potential to offer a company large savings. By building a prototype of such a system prior to a major resource commitment, the risk of failure and the associated cost would be drastically reduced.

To achieve the maximum return on a prototyping investment, the risks associated with systems even a fourth the size of the hockey example must be controlled and reduced. In order to manage the inherent structural complexity of such problems, the general ability of the data base manager to effectively model any structure must carefully be considered.

The functionality of the data base manager is an important contributor to the effectiveness of the prototyping effort. Its suitability requires judgment and consideration of context; its overall integration with the other components.

12 *The prototyping software must be conceptually consistent across components to permit virtuoso prototypers.*

As was discussed in the section on staffing (Section 4.1), prototyping must be done by small teams. As a consequence, the prototypers must be able to execute all the software components. The prototyper, in actuality, must be a "one person programming band" (Figure 4.14).

The depth of knowledge required to maintain the dictionary, develop screens, generate programs, define data bases, generate documentation, and so on, must not require the availability of subject matter experts.

To accomplish this, the product must not only be internally integrated in terms of sharing the dictionary, but it also must be externally integrated in presentation to the prototyper. Syntax, terminology, error message presentation, help facilities, function key assignments, and so on, should be as consistent as possible across components.

13 *The prototyping software should provide security features.*

For many applications, security is an important requirement. The common techniques for enforcing security,

Figure 4.14 One person programming band. Each prototyper must be able to manage all software components as a virtuoso performer.

□ terminal level
□ user password
□ transaction level
□ data encryption
□ record level
□ data element level
□ read only access mode

should be provided by the software.

14 *The screen generator should simplify the creation and maintenance of screen maps.*

Many structured applications tend to be full screen oriented. The ability to build and modify screens quickly is important. The screen program should also be able to do "automatic editing" of input fields per declared specification. Screens that fail "automatic editing" should automatically be looped with the terminal operator until all fields are valid.

The function delivered by a screen generator should include the following:

The screen generator should permit interactive painting of the screen format.

The screen generator should provide the ability to move a field by changing its vertical and/or horizontal coordinate.

The screen generator should provide the ability to specify the characteristics of selected groups of fields at a global level.

The screen generator should provide the ability to associate edit-rule tables and messages with a field. The failure of a field to meet an edit rule would result in "automatic editing" displaying the associated message on the screen.

The generator should have a test driver to permit testing the screen without creating a program.

The generator should be able to copy an existing screen as a starting point for work.

An external mask (i.e., ZZZ,ZZZ.99) should be able to be declared for a field. The generator should automatically edit and convert the field from the mask to a standard format.

While executing the screen generator, the dictionary should be accessible to check definitions.

These requirements are not meant to be inclusive but to make a point. The screen generator should be functionally powerful. Much of the rote "tunnel" editing that has traditionally been done procedurally is suitable to be done automatically by declaring the necessary definition in a screen program.

15 *Most of the conventional "tunnel" editing rules should be definable by declaration.*

Many of the "tunnel" edits that compose the "model" edits, which were discussed under Prototyping Principles, Section 3.3, are fairly standardized and should be done automatically by the prototyping software. Edit-rule tables should be definable to the dictionary that specify edits such as:

- ☐ required/not required
- ☐ within (not within) contiguous or noncontiguous table ranges
- ☐ pattern matching
- ☐ minimum number of input characters
- ☐ contains (not contains) character string
- ☐ alphatetic (nonalphabetic)
- ☐ right/left justification
- ☐ reordering of characters
- ☐ encoding and decoding (translation) of data elements
- ☐ date checking by various formats

Such an edit-rule table may also include the ability to do some simple procedural logic; checks such as an inputted date is prior to the system date.

The goal of the rule tables is to off load "tunnel" editing to the prototyping software. By associating the rule tables with elements on the screens, the editing required to be done procedurally is minimized.

16 *The procedural language component of the prototyping software should enforce the concept of structured programming.*

Most conventional programming languages permit the creation of "spaghetti" code. Uncontrolled branching between program segments is permitted. It is the generally accepted industry consensus that "structured" programs are preferable. A structured program is composed of only four primitive constructs:

- ☐ Sequential
- ☐ Select
- ☐ Do Until
- ☐ If then Else

Any necessary program logic can be constructed by appropriate combinations and nesting of these constructs.

The procedural language component should only syntactically permit the creation of structured programs. Unstructured constructs such as the COBAL ALTER or GO TO should not be part of the language.

17 *The prototyping software should "bind" as late as possible in the creation process.*

Binding is the act by which the source definition of entities become executable entities. Maximum flexibility is achieved by all binding occurring dynamically at execution. Languages that bind at execution are often referred to as interpretive.

The advantage of late binding is that any change is automatically rippled through all assembled entities that use the entity. Obviously, the price one pays for execution binding is performance, though this would be inconsequential to prototyping software. Since the software must be selected from production oriented software it is a practical consideration to remember when analyzing candidates.

18 *The prototyping software should have extensive debugging facilities.*

Prototypers, like any other developer, will make errors. Though much can be off loaded to the prototyping software by declarative specification, procedural code will still have to be written. Bugs are an unfortunate reality of programming.

The prototyping software should have extensive facilities to aid in the detection of execution errors. Facilities such as the following would be desirable:

- ☐ trace facility
- ☐ interactive debugger that permits the dynamic setting of break-points, displaying of field values, and so on
- ☐ test data generator
- ☐ trapping of abends with display of offending source statement and variables
- ☐ interactive file print utilities

The prototyper should not have to deal with machine level debugging procedures. Debugging facilities should all be operable at the source code level. Reference to core dumps, internal register contents, internal system control blocks, and so on, should not be required.

19   *There should be a complete set of training and documentation facilities to support the use of the software.*

The vendor must be able to support training and provide accurate documentation on how to use all software components. Training classes should include case study problems and actual laboratory hands-on experience. Documentation should be available in multiple packaging:

- ☐ complete component documentation (conventionally in large binders)
- ☐ quick reference cards
- ☐ interactive run time help
- ☐ topical index that cross-references all documentation items

A prototyper will tend to have breadth of skill but limited depth due to the multiplicity of products that must be used. The prototyper's productivity is consequently amplified by the availability of good and accessible documentation.

20 *The prototyping software should support the "versioning" of the prototype.*

During iteration, baseline configurations will be established. As a consequence of the demonstrations, possibly extensive modifications will have to be made to the baseline. The software should support the creation of multiple versions of an application entity. Reversion may be necessary. The problem is one of release management. Multiple releases must be able to be stored concurrently with appropriate facilities to manage movement between them. Progress in prototyping can sometimes be two steps forward, one step back.

21 *The prototyping software should permit the presentation of logical records to the application programs.*

It is advantageous that the physical structure of the data base be as hidden as possible from the application programs. During the iteration steps, elements, records, and inter-record relationships will undergo revision. A logical record interface between the application programs and the physical data base would shield existing programs that are not logically impacted by the change. Only those programs that need to use the changed structure would have to be revised.

The concept of an "invisible" data base structure is illustrated in Figure 4.15. The data base view component extracts from the physical data base only the subgroup of data elements needed by the program into a logical record. The data base view facility effectively shields the application from any future changes to the data base structure by creating a highly program cohesive record. Even when and if the physical data base is changed, the data base view will continue to deliver the same logical record to the program.

The impact of data base change is localized and constrained to only those parts of the existing prototype that are functionally effected. Obviously, dealing with logical as opposed to physical records markedly improves the flexibility of revising the prototype. Only the directly impacted areas need to be revised. If physical records are being used by the programs, all the programs, regardless of actual need, will have to be revised. The logical record concept provides an extremely high productivity method to maintain independence between the physical data base and the application programs.

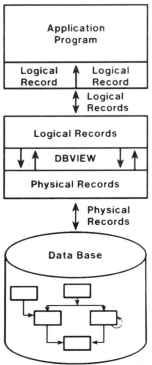

Figure 4.15 Logical records. The applications programs are insulated from the actual physical data base structure.

Note: Obviously, relational data base managers provide a great deal of this type of functionality automatically.

22   *The prototyping software should permit the dynamic run time (application program) declaration and manipulation of simple file structures.*

The definition of a data base structure and associated data base views can in some instances be both too time consuming and overkill for the problem. The data base manager should have the ability to permit the run-time definition and manipulation of records into and out of a data base.

The storage area for such a dynamic data base may be placed in another software component such as the dictionary and would generally be referred to as a queue area, scratch area, or program work area. The types of record manipulation capability required for such records should be synonymous with a predefined data base record:

Declare a valid record type.

Add a record.

Retrieve a record by key.

Revise a record.

Delete a record.

Sequentially retrieve the next/prior record.

The records should stay in the data base until explicitly deleted.

The records should be retrievable by all software components.

This type of facility permits the prototyper in simple or rushed situations to simulate procedurally a relational data base structure without the need to explicitly do any data base definition. Each "relation" equates to a declared record. Programmatically, all the required logic to insure inter-record key integrity and mimic a JOIN operation could be accomplished. Though certainly not an appealing alternative for dealing with large and complicated data bases, the raw availability of such function increases the resources and choices available to the prototypers to respond in proportion to the needs of each problem.

Software Requirements Summary

These requirements are not inclusive and certainly others, especially at the most detailed level, could be suggested. However, they clearly set the requirements tone. The thrust of all requirements is to maximize the slowest and most error prone component of the building process, the human prototyper. Any function that could be performed by nonprocedural declaration regardless of potential performance penalties is desirable to procedural effort. Off-loading effort to the most intelligent software is the best strategy for an optimum productivity environment. The prototypers can only work at a human pace. They must work much smarter but not harder.

Traditional software evaluation criteria such as

☐ recovery/restart capabilities
☐ data base size limits

□   terminal network size
□   completeness of utilities

are a secondary consideration after reviewing the prior factors. It is extremely doubtful that a prototype will even come close to the limits of a product created by a reputable firm. Since the existing choices are in fact production oriented architectures, a two terminal prototype using a four track data base and one printer will hardly be noticeable.

This discussion brings back the question of whether the prototype should or could become the production system. If the strategy is that it should, then the production center features of the software become quite important. If the prototype is to be refined into the production system, then the software architecture must effectively support the requirements of both centers.

Though this is certainly desirable in some situations, the overriding issue in selecting the software must be its prototyping functionality. Model airplanes that are put into wind tunnels are not ever expected to carry passengers. Miniature houses that are built to demonstrate an architect's interpretation of a client's needs are not meant to be lived in. Philosophically, one should accept the prototype only as a model and not attempt to cheat the process. A great deal of service is delivered to the development process by the creation of models. When the prototype creation process has to serve two masters, neither may come out very well.

The prototyping software selection process should be focused on meeting the needs of the prototypers. Target architectures for implementing the prototype, including the prototype architecture, should be evaluated after prototype acceptance based on cost and technical feasibility. The creation of an accurate and vivid requirements document is sufficient justification for building models without the need to have implementation potential.

The requirements that have been listed are met to varying degrees by commercial products which exist today. Though some compromise of requirements or in-house extension of the product might be required, careful consideration of modern data dictionary driven architectures will yield a satisfactory candidate. It is hoped that in the near future as the market for prototyping software grows, products specifically designed to meet the needs of prototyping will become available.

4.4 THE WORK ATMOSPHERE

Enabling the prototyping process requires the creation of a work atmosphere that directly contributes to exceptional productivity. Neither "tinker-toy" software nor talented prototypers will be able to compensate for response time measured in minutes rather than seconds or wasted time queuing up for access to a terminal. There are specific ways to structure the work environment to remove productivity inhibitors and promote the overall prototyping effort.

There are seven specific actions that can be taken to enable the process:

Project Work Rooms.   Self-contained facilities for developing the models.

Response Time.   A consistent and high level of interactive and batch response.

Repeatable Construction Process.   The definition of a set of manual procedures for organizing and executing the manual parts of the building process.

Library Resources.   The availability of both technical and business libraries to the prototypers.

Demonstration/Presentation Facilities.   The availability of presentation facilities that make the demonstration process comfortable and convenient for all reviewers.

Centralized or Distributed Prototyping Centers.   An organizational strategy as to whether to provide a single central resource or to make prototyping an extension of each development department.

Parts Department.   A prototyping support function that certifies and catalogs potentially high reusable parts which are application independent.

How you respond to these work environment issues can make the prototyping simpler or harder.

4.4.1 Project Workrooms

A project workroom is a self-enclosed facility that has all the necessary resources to perform the development task. Terminals, reference manuals, work desks, easels, and so on, for the project team are all contained in one cohesive work place.

A project workroom has many advantages. The close working environment encourages synergy. The team members will begin to compliment each other. Communication problems are minimized. No appointments have to be made. Instant walk throughs, inspections, and reviews are possible. All the necessary resources are defined and available in one place. Manuals do not have to be searched for nor do terminals have to be shared. If the prototypers are to be held accountable for the rapid delivery of a model, they must have control of their resources.

In addition to the customary and routine items you would expect in a workroom, there are four mandatory items:

Interactive Terminals.   These are the prototyper work benches.

Batch Print Terminal.   This is the remote printing terminal for batch printouts.

Reference Library.   This is a complete set of software documentation.

Visual Aids.   These are eye level aids to facilitate recalling detail about the software or prototype.

Interactive Terminals

A one-to-one ratio should exist between prototypers and interactive terminals. The terminal is the workbench. Access should be available as required. To accrue the benefits of a workbench tactic fully, a workbench must in fact exist for each prototyper.

A supplementary "hot" terminal always on line to the data dictionary should also be considered. In a dictionary driven environment, the dictionary is the only credible source for documentation about the model. Dictionary driven architectures can provide a paperless development environment. Access, as needed, to the dictionary should be enabled. A "hot" dictionary terminal permits reviewing dictionary records while in the middle of another software component without having to abort or suspend the effort.

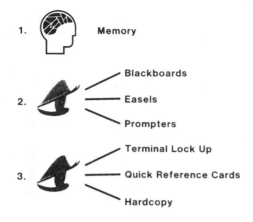

1. Memory

2. Blackboards
    Easels
    Prompters

3. Terminal Lock Up
    Quick Reference Cards
    Hardcopy

Figure 4.16 Workroom recall aids. By careful design of the workroom, the prototyper's memory can be amplified.

Batch Print Terminal

A medium to high speed printer should be available for the local routing and printing of batch runs. Such a printer may be shared between project rooms but should be in immediate proximity. Quick access to the results of batch runs that need to be printed rather than browsed on the terminal is almost as important as interactive access.

Reference Library

A complete and current set of reference documentation on how to use each software component should be available in the workroom.

Visual Aids

While building the prototype, the developers need to "remember" many specific things. The rules for controlling each software component, the data base structure, the names of elements, programs, records, and so on, all have to be recalled as needed. Obviously, nobody will retain all the detail in the most efficient recall mechanism, their memory. Consequently, it is advantageous to arrange the prototyping room to aid the recall process (Figure 4.16).

Aid: Wall-to-Wall Blackboards

It is extremely beneficial to line the project room walls with blackboards. Blackboards provide an excellent scratch tablet for testing ideas and for documenting primary system entities at eye retrieval level. By simply looking around the workroom, recall access to numerous critical information about the model can be accomplished. For example, blackboards could be sectionalized and contain the following types of information:

- [ ] a graphical drawing of the data base structure with the record and record key names
- [ ] a "completed work" board with the names of each completed program and associated screen
- [ ] a "componentry board" with the names of system entities that are high reuse or "cut and paste" candidates
- [ ] a "record name" board with the names of all work records available for the system

The blackboards can serve as an eye level global index into the prototype.

Aid: Personal Easels

If the blackboards can act as global memory aids, easels can serve as local (personal) aids. Each prototyper can document on his or her own easel notes or definitions that which is of a personal and immediate value. As she moves to the next function, the easel paper is easily ripped off and a new set of immediate and temporary notes can be jotted down. The eye level indexing support is distributed to support the individual's needs as well as the global project needs.

Aid: Prompters

There is information of high recall value that crosses projects. Prompter charts that itemize function key assignments of each software component or summarize prototyping standards can hang down from the ceiling. Having the function key assignments at eye level can reduce mistakes. Having the standards at eye recall level, can provide sufficient information to locate a needed entity. For example, suppose a center standard is that for each data base record, three work records will be created in the dictionary with the suffixes -work1, -work2, and -work3 appended to the record name. Recall of this standard when combined with viewing the data base blackboard identifies the names of any needed work records.

If the visual aids are insufficient in providing the level of detail needed, less productive recall mechanisms can be used.

Aid: Interactive Lookup

The dictionary and software help functions can provide rapid retrieval of necessary information. Since the application is completely defined within the dictionary, by requesting the display of a named entity, all the attributes and relationships of that entity can be reviewed. Similarly, interactive help facilities can provide syntax and usage rules for each software product.

The blackboards, easels, and prompters can play a vital role in enabling interactive recall. All the entity names, records, programs, and so on, on the blackboards and easels are the direct access keys for the dictionary. They provide an eye level index to dictionary detail.

Aid: Quick Reference Cards

Quick reference cards are a convenient mechanism for jolting one's memory. They normally provide a syntactical summary of a product without the narrative detail. Such aids can be referenced for each software product and center standards.

Aid: Hardcopy Lookup

The slowest way to recall needed information is to have to browse documentation binders or printouts. Of course, this is necessary if total detail is needed or extensive comparison and analysis is being done.

By carefully designing the work room environment, productivity can be markedly influenced. Every time a necessary recall is accomplished by "looking up" rather than browsing a hardcopy printout, productivity is enhanced. Figure 4.17 portrays what a workroom might look like.

4.4.2 Response Time

There is little point in conducting a prototyping strategy if adequate response time for interactive and batch jobs is not provided for. Having raised the user's expectations, and selected a talented staff, it is contrary to the entire thrust of the effort to retard the building process by inadequate response time. A simple guideline would be as follows: Interactive

Figure 4.17 A project workroom. The composite workroom enables high prototyper productivity.

response time at the high level should be 5 seconds; batch turnaround should be in the vicinity of 15 minutes.

Rapid response takes on a very important role during demonstrations. People tend to bore easily and lose interest and confidence if the minutes tick by while waiting for a screen to appear. Good response, consequently, is not only needed to expedite the demonstration process but to maintain user satisfaction with the entire process.

As for the building and refining process, it is obvious that no response equates to no work being accomplished. The frustration level achieved by prototypers ready and able to work is tremendous. It is unreasonable to hold a prototyping team accountable for delivering a model, if one of the most fundamental needed resources, adequate machine access, is denied.

4.4.3   Repeatable Construction Process

As has been alluded to before in the preceding chapters, there is a clear distinction of perspective of the prototype between the user and the prototyper. The user views the prototype as a system in motion at a temporary baseline. They, by the clear intent of the process, are urged to refine and evolve their needs. The prototyper, though, has perpetuated an illusion. Though the user may feel and act as though she or he is dealing with "play-dough," the prototyper still has to manage and manipulate physical programs, elements, records, and so on. Though it is a prototype, it is, in actuality, a small system with all the associated potential problems in maintaining and enhancing it.

The integrated dictionary driven architecture that was itemized as a mandatory requirement to support properly the prototyping effort addresses many of the problems. However, just as user applications are composed of two parts, an automated computer subsystem and a manual human subsystem, the construction and evolution of the prototype also has two subsystems.

In this case, the selected prototyping software composes the computer subsystem and the set of procedures and standards used by the prototypers to interface to the software composes the human subsystem. The predictability and efficiency of the human subsystem is a critical contributor to the overall productivity effort.

The manual procedures involved in constructing the prototype should be based on a defined and repeatable process. This requires conventions and standards for performing the repetitive human functions. Just as it is wasteful to rewrite a program whose function is already defined, tested, and available in the dictionary, it is unproductive to commence each prototyping project without preestablished procedures for conducting the effort. Since the construction process is repeatable across prototypes, experience should yield a standard set of procedures for organizing the effort.

Discipline and rigor are appropriate for the construction phases of the SDLC. Heuristics are appropriate for soliciting requirements. The actual act of building the prototype is a construction problem. It is best done by conducting oneself with discipline. Discipline, however, does not equate to predocumented formality.

Standards are very important to this effort. They permit unambiguous communication and minimize clerical decision making. Entity names can be automatically determined by rote. This enables the locating of existing system entities. A dictionary will become a hodgepodge of chaotically named entities if specific rules are not adhered to in their naming.

Consider the following as part of a set of naming standards:

System Identification (ID). This is a four position alphabetic code uniquely identifying a project.

Program Name. This is an eight position code of the format *XXXXPNNN* where *XXXX* is the system ID and *NNN* is a sequentially assigned number.

Map Name. This is an eight position code of the format *XXXXMNNN* where *XXXX* is the system ID and *NNN* is a sequentially assigned number.

Message Number. This is an eight position code of the format *XXXXSNNN* where *XXXX* is the system ID and *NNN* is a sequentially assigned number.

Though these standards are certainly simple and unoriginal, they deliver a number of very productive results:

Procedurally, at the beginning of each prototype, a unique application tag is assigned to each application.

No decision has to be made on how to name any entity.

No guess has to be made on how any system entity was named.

Communication between prototyping members can take place within a tight and precise jargon.

All the entities that comprise a prototype are pretagged for convenient grouping for documentation.

The physical prototype has a naming structure that is easily learnable by nonprototypers.

Procedurally, a "bakery ticket support system" is required to administer the assignment and tracking of "next names." Such an application might also record the date and person who selected the name. Of course, administration of standard names is universal across all prototypes and should be part of the repeatable process.

The translation of the prototype to the production system is expedited by the use of standards. A prototype as a specification, must be read, analyzed, and interpreted. The ultimate implementors will reject a set of dictionary reports that reflect ad hoc conventions. A prototype is not a personal system and must be constructed with the intent to communicate with all its users, developers as well as end users.

A set of repeatable procedures and standards for conducting the human interface to the prototypes also aids in training the next generation of prototypers. Once a new member of the prototyping center learns how one prototype is organized, she or he essentially knows how all prototypes are organized. The training of new staff has been simplified to one software architecture complimented by a standard human interface.

4.4.4   Library Resources

The prototyping center should have access to both technical and business libraries. There is a substantial amount of books, magazines, periodicals, and journals on many of the subject applications that will undergo prototyping. Many people have already solved similar problems and have documented their solutions. The literature can provide at minimum introductory information on the topical business area and at best provide an existing solution.

The "user needs" step of the prototype life cycle should not be exited until the prototypers can explain the application back to the user. One of the ways to increase comprehension is to review general literature on the subject area. In essence, the library resource can compliment the user's explanation of the business function by providing additional perspective.

The same reasoning holds true for the technical aspects of prototyping. The prototyping center as part of its mission must strive continually to "improve the process." Improving the process not only includes revising procedures based on your own experiences but also identifying what others are doing and evaluating it for possible use. The technical and procedural problems associated with building and refining application models will not be unique to each organization. At worst, identifying what others have done wrong will prevent us from making the same mistake. At best, an improved method for perhaps packaging the hardcopy documentation may be suggested. Only by actively following the technical literature will we be able to leverage the experiences of our peers at other companies to our benefit.

4.4.5 Demonstration/Presentation Facilities

A prototype has to be reviewed multiple times by numerous people before the functionality is agreed to and a consensus approval is reached. Sitting in front of a CRT screen for extended periods of time is often tiring and is impractical if the number of reviewers exceeds two. People will not be effective in analyzing the prototype if they are squinting their eyes or jammed for space by other reviewers.

It is important to the success of the iteration process that it accommodates the reviewers and encourages participation. The demonstration process, consequently, must be comfortable for the reviewers. The best way to accomplish this is to have the display terminals attached to large screen projectors. The screen image portrayed on a large screen projector can be viewed comfortably by a substantial audience without eyestrain or the cramped proximity of other reviewers. The reviewers can sit at work tables with notes spread out. Any screen can be discussed as long as necessary in a comfortable setting.

The practical reality is that as the applications being prototyped get larger, the word "user" becomes very plural. There are many users who need to critique the model. Often, the only way to review certain functions is by having a large group of users concurrently. Consensus requires group consideration of the model. Normal CRT screens are simply ineffective for performing such demonstrations.

The requirement is then as follows: It is desirable and mandatory to be able to demonstrate the prototype to various size groups of reviewers concurrently. The demonstration process must be comfortable and facilitate active participation. It is hoped that by centering group discussion on the sample screens, innovative ideas and suggestions for improved functionality will be generated. A large screen projector in a setting of tables and chairs best meets these requirements. The audience can be as large or small as required and extended reviews of the prototype can take place without straining the reviewers.

4.4.6 Centralized or Distributed Prototyping Centers

In a small company, there is no question as to whether to centralize the prototyping service or to distribute it. As there is probably only one analysis/development group, the placement of the prototyping staff is straightforward. In a larger organization, the question is more difficult to answer and requires some consideration.

A prototyping center could be established as a central resource to all development organizations. As requested and required, project managers could use the prototypers as project consultants. The advantage of such an arrangement would be high resource efficiency. All the prototyping resources could be optimized and apportioned for a steady and predictable work flow.

A prototyping center could also be established as a distributed resource where each major development area, marketing system services, treasury system services, accounting system services, and so on, has its own center. Each prototyping staff would be targeted to serve a specific business function. This is potentially a highly effective approach since the staff will become highly familiar with the business area. Unfortunately, efficiency may be compromised since the work flow between business areas may be inconsistent making consistent staffing and utilization difficult.

Which way is better is dependent on the size of the organization and the management style. My own preference is a distributed approach. Centralized functions tend to be resource efficient at the price of reduced effectiveness in responding to problems. Distributed functions tend to be more effective in responding to and servicing needs but at the price of redundant staffing and resources.

Given the possible tradeoffs between accenting efficiency or effectiveness, I would vote for effectiveness. The life cycle benefits to be accrued by building models will more than offset any minor resource inefficiencies. Proximity of the prototypers to the user via the responsible development organization has clear advantages:

The prototypers will become expert in the subject area and minimize the learning curve for each new problem.

The prototypers will develop a consistent working relationship with the user organization.

The politics of having prototypers from one organization work with developers of another organization will be eliminated.

The distribution of the prototyping function to multiple development organizations does not necessarily require a physical distribution of the facilities. A single physical center could be shared by multiple development organizations. This approach could help alleviate efficiency issues.

4.4.7   Parts Department

Reusability of existing components has to be promoted at two levels: locally with application only sensitive parts and globally with parts that are generic in function. A parts department serves the function of providing global parts to all prototypers. The functions performed by a parts department are as follows:

Selection (creation) of parts that have a high global reusability potential.

Certification of correctness.

Cataloging of parts to permit easy and quick accessing by the prototypers.

Parts can be both finished components or skeleton solutions. A set of date conversion modules is certainly a high reusable and finished part. A module could alternately contain a skeleton logic structure for browsing a record types. Parts could be as small as a message: "PLEASE ENTER NEXT FUNCTION" to an entire subsystem such as HELP. Figure 4.18 illustrates the concept of automated parts administration.

Unlike a manufacturing environment where a major problem with parts administration is stocking and maintaining inventory, the primary problem with parts administration in this environment is the identification by the prototyper of a part. The parts department needs to provide a sophisticated locator service to encourage and enable the selection of global parts. Such an indexing system must provide multiple key word accessing strategies and extensive browse capabilities. It must be both simple and quick for the prototypers to find what they need.

Parts administration is not a novel idea and has certainly been attempted with mixed results in most conventional environments. One should anticipate exceptionally good results in a prototyping situation for two reasons:

The prototypers are component engineering/reusability oriented.

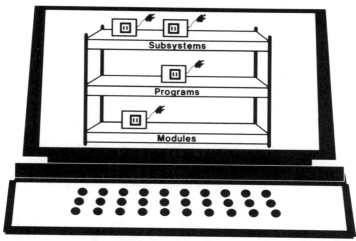

Figure 4.18 Parts administration. A parts department should support the prototyping effort by identifying, certifying, and indexing software parts with a high degree of reusability.

The prototyping architecture supports component engineering/reusability.

Both of these are often missing in conventional situations.

4.5   PROTOTYPING CENTER SUMMARY

Each of the resources required by a prototyping center makes its own special contribution to the process. Yet it is really the integration of them all that makes the significant difference. The whole is indeed much greater than the sum of the parts.

The staffing is obviously the most critical factor. The high visibility and accountability of the activity requires competent and able performance. Still, the assignment is challenging and cannot be done by simple will and brute force.

The prototypers, however competent, will be limited in meeting the goals and objectives of prototyping if the software/hardware architecture used is inappropriate. Though the software can be compromised to deal with the specific types of applications you envision using prototyping for,

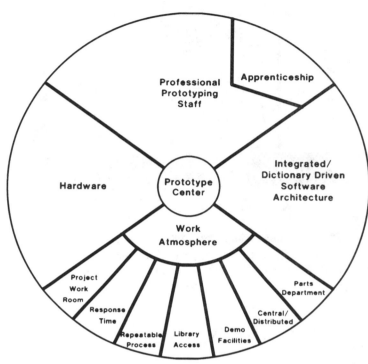

Figure 4.19 The composite prototyping center. The complete prototyping center consists of staff, hardware, software, and work atmosphere.

in general the software requirements that were listed are mandatory. Consider the consequences of attempting to do a problem the size of the "hockey data base" without them. If the software is not self-contained and integrated, the labor intensiveness and error proneness of redefinition will retard any progress. If you cannot reuse parts or quickly develop new parts from existing parts, the volume of work will drown you. If you have to code all the editing rules procedurally as opposed to declaring them, you will bog down in insignificant detail. If the dictionary cannot completely and accurately document the system, any iteration cycles will be extremely susceptible to undesirable error rippling. If the data base manager cannot directly model the problem, unproductive effort will be required to trick the data base. Artificial design gimmicks will only come back to haunt you later. The selected software is clearly the engine that moves the process forward.

Finally, the work atmosphere completes the center requirements. Project workrooms provide a high intensity focal point for the building activity. The availability of large screen projectors for demonstrations assure the attentiveness and careful consideration of all reviewers. A decision as to whether to centralize or distribute the prototyping function can provide flexibility in balancing efficiency and effectiveness. A parts department provides the potential for the ideal prototype situation, the construction of an entire prototype almost completely from existing parts. Figure 4.19 summarizes the resources of a prototyping center.

FIVE

PROTOTYPING AND PROJECT MANAGEMENT

Prototyping does not occur as an isolated event. It is a subprocess of the greater development activity and as such it comes under the control umbrella of the project management function. In general, the project management function is responsible for managing five aspects of a project: quality, resources, cost, time, and technology. The planning, control, and organizational activities performed by the project management organization to control these five project attributes make an important contribution to the overall success of the effort.

The introduction of prototyping results in the need to revise certain project management procedures. Many of the current practices that have been adopted by modern project management organizations originated in response to the same problems which resulted in rigorous definition. By applying careful planning, properly organizing the effort, and controlling the effort through systematic tracking of actuals against the plan, unpleasant surprises would be avoided.

Prototyping does not alter the validity and appropriateness of applying sound project management practices to the overall project. It does, however, require some modifications. In response to the historical development problem, project management techniques have tended to favor extreme control. The prototyping phase requires a reasonable amount of looseness and flexibility.

5.1   CONFIGURATION MANAGEMENT

Consider, for example, the project management technique of configuration management (CM). The thrust of CM is as follows:

Define a system deliverable.

Rigorously control all changes to that deliverable.

Reconcile at implementation that the final product is consistent with the original definition as adjusted by the approved changes.

A system deliverable that is brought under CM control is referred to as a complete program configuration item (CPCI). Baselines are established for each CPCI as follows:

Functional Baseline.   This is a general description of the deliverable.

Allocated Baseline.   This is a detailed definition of the deliverable.

Product Baseline.   This is a description of the operational CPCI.

A change to any baseline document may be proposed by any project member. A change proposal document is called an engineering change proposal (ECP) or if an operational defect is discovered it might be referred to as a software trouble report (STR). The ECP/STR explains the reason and nature of the change. Each ECP/STR goes before a system change control board (SCCB). The function of the SCCB is to judge the validity of the change request and, if adopted, assesses the impact on all CPCIs.

A CM staff tracks the status of each CPCI, ECP, and STR. The staff insures that all CPCI baselines are maintained in synchronization with only approved ECP/STR. If necessary, audits may be made to insure consistency.

Though one cannot possibly argue with the logic and precision of this project management technique to control change, it is a good example of the types of project management techniques that are inappropriate for prototyping. Techniques of this rigor and formality will squash the spontaneity, speed, and flexibility of the whole prototyping process.

5.2   FLEXIBLE CONTROL

Elaborate project management control is not appropriate for the prototyping step. Project management is needed to enable the process but not to stifle it. Each prototyping cycle is relatively quick. There is clear responsibility, accountability, and visibility of the process. It is obvious if the prototypers do not deliver a model. It is also obvious if the user does not attend review sessions. A status report can be provided by a demonstration of what is currently working.

The risk of limited control is compensated for by the small prototyping teams and short delivery cycles. An iteration cycle involving two prototypers for a 2-week period will deliver at worse a minor failure. Since the roles are well defined, the product extremely visible, and the delivery periods are in short intervals, a loosen project management approach can be adopted without incurring substantial risk.

Though the grip of the project management function should be loosened, it certainly should not be eliminated. The project managers need to participate in the initial decision as to whether prototyping is an appropriate strategy for the application. Meetings and demonstrations with users have to be scheduled and planned for. The details of the hand-off documentation have to be negotiated.

Basically, the project management function needs to perform its normal tasks, but modify the formality to allow for the special nature of prototyping. Formal project management procedures such as typified by CM would defeat the entire thrust of the prototyping process. It would discourage change and retard the rapidity of the process. Speed and formal bureaucracies are incompatible. The prototyping process will self-correct itself during iteration if a major mistake is made. It is philosophically better to do something and experience it than to debate it.

5.3   MAJOR PROJECT MANAGEMENT ISSUES

There are four specific project management issues that are impacted by prototyping:

Estimation Procedures.   These are the methods used to estimate both the time/cost of the prototype and the target system.

Chargeback.   These are the procedures used to allocate prototyping expenses back to the users.

Change Control.   These are the procedures used to approve and control changes to the model as a result of the review process.

Sign-Off.   This is the acceptance of the prototype by the user.

Each will now be reviewed for its impact.

5.3.1   Estimating Procedures

Estimating the time and cost for a project is a difficult task. It often takes on the characteristics of negotiations and compromises rather than an actual independent estimate. Since traditional estimating is as much a political problem as a technical one, it would be advantageous to simplify the estimation process for the prototypers.

The requirements for a simple estimating algorithm are straightforward:

Initial Build Step.   The time estimated must be sufficient to allow a meaningful model to be developed, maintain user interest and excitement, and negate the risk of an expensive throwaway.

Revision Step.   The time estimated must be sufficient to maintain project momentum, allow for meaningful change, and maintain quality.

Both Initial Build and Revision Steps.   The time estimated must be flexible enough to deal with the variations in project size and complexity.

Experience indicates that an initial model should be deliverable by a two person prototyping team within a 2–6 week period following the completion of the user needs analysis. The primary factor affecting the estimate is the size and complexity of the problem.

Experience also indicates that revisions to a model should be deliverable between instantaneously, in front of the user, and 2 weeks. Early iterations where major functionality enhancements are developed would tend to require more time. Later iterations where fields are merely being relocated on a screen would tend to be done immediately.

These estimating guidelines meet the requirements and eliminate the politics from the estimating process. They are long enough to permit meaningful work but short enough to prevent a major directional error. They constrain the amount of work that can be done per iteration and permit the demonstration steps to control the quality of the model.

The prototyping process also helps solidify the estimation of the cost of the real system. The prototype is the "document" against which the real system is to be estimated. Since the deliverables are well defined, it should be substantially easier for the implementation team to assess the cost of delivering the final system.

The implication of prototyping on cost estimating is that project management should invoke a two-step estimating process for costing a system. First, the user should pay for time and materials to develop the requirements. The cost is highly variable and depends on the dynamics of the iteration cycles. After a prototype is agreed to, a solid estimate of the development costs can now be made.

The problems of delivering meaningful estimates are markedly reduced. No one can predict the time to develop requirements since they are an unknown. The realistic answer is: It depends. Once a solid definition is obtained, it is realistic to expect a firm commitment of time and cost since the change and looping that historically destroys estimates have been minimized.

5.3.2 Chargeback

All the expenses incurred in developing the model should be billed back to the user organization. Prototyping incurs labor and machine expenses. A user who is paying the bill will tend to take the process seriously with a careful eye for efficiency.

Chargeback is the most efficient mechanism to control the iteration cycle. Since the iterations cost money, the user will have to balance additional function against cost. If prototyping expense is spread against all users, this important control mechanism on iteration would be lost.

5.3.3   Change Control

Though the CM example that was cited as a means to control change is excessively rigid for prototyping, it does raise a critical issue. Who decides what changes should be made to the prototype? It is fine to talk in generalities about the user dictating numerous refinements based on the demonstrations but the reality is that user is often a plural and multidepartment entity.

Do all the users have equal authority to request changes? Is a plural consensus required before changes are made? When the word user does not equate to a single decision maker, change procedure is a practical problem.

The most appropriate way to approach this problem is for the project management organization to originate an interactive control board. A small group of designated individuals have to be present at all demonstrations to approve or disapprove changes. Someone has to have clear decision-making authority.

This is neither as powerful nor ominous as it sounds. The iteration process can rectify poor decisions. If a suggestion doesn't work out, it can be removed. If a reviewer persists in her or his suggestion, it can be tried in a later iteration. Decisions, however, have to be made quickly to maintain the momentum of the process.

5.3.4   Sign-Off

In a traditional definition approach, sign-off by the user is based on the descriptive/graphical model. In a prototyping environment, sign-off equates to approval of the demonstrated prototype as the desired system. The user is agreeing to a firm baseline. What he or she has seen and experienced is in fact what he or she should get.

There are multiple implications from this type of sign-off:

The project manager should invoke a rigid change control process to insure delivery of the user accepted prototype.

The user has responsibility for the impact of any changes that will alter the system delivery.

The developer is committed to copy the external presentation of the prototype into the operational system.

If these responsibilities are not adhered to, the entire purpose of the prototyping will be lost. Responsibility for developing the definition has been placed where it should be on the user. It is now everybody's responsibility to implement that vision. Only the user should change that vision and in so doing take responsibility for the delivery consequences.

This type of discipline is required to ensure that the model building is taken seriously, and a truly representative and ultimately acceptable model is developed. Business will no longer proceed as normal. Prototyping provides the opportunity to throw one away. Careful and considered advantage of the opportunity should be taken by all project members. If you don't take the process seriously, reality will be a mean taskmaster at implementation but this time there will be nobody else to blame.

5.4 SUMMARY

Project management is an important contributor to the overall success of the system development effort. Just as prototyping causes a redefinition in roles and procedures for the users, it requires a revision of technique on the part of the project managers. Though the basic tenets of project management are still valid and need to be applied, the tactical implementation of the project management functions have to be loosened to permit the prototyping process to work effectively.

Prototyping cannot work under an extremely tight project management approach. Rigorous project management approaches such as CM are too constraining on the process. The flexibility, responsiveness, and robustness of the prototype cycle is completely lost when formal documentation and approval procedures are required before each move.

To compensate for the perceived weakening of project control, prototyping provides high visibility and accountability. Risk of a prototype out of control is constrained by the short periods between model deliveries. It is obvious to all what is occurring. Relative to the cost of correcting errors in the traditional SDLC, throwing away a 2-week effort is a nominal expense.

Project management needs to be redefined but not abandoned. Prototyping is not an excuse to avoid proper management control of the development effort. It merely has to evolve to optimize its contribution in a new situation.

SIX

INTRODUCING APPLICATION PROTOTYPING

To accrue all the benefits that have been discussed, it is necessary that prototyping be offered as a formal and standard service by the development organization. The introduction of a prototyping methodology into an established environment is a technology transfer problem. A new technology, in the form of the composite prototyping center, has to be made part of the normal and preferred practices of the organization.

Some people automatically resist any change. Many people will legitimately be skeptical of the productivity claims given the unfulfilled promises of prior techniques. Both individual and group attitudes will have to be changed. Unless the organization is unusually open-minded and progressive, a strong business case will have to be made and many people will have to be convinced that the claims are indeed accurate.

The best way to introduce an organization to prototyping is by a structured and planned project. The goals of the project are to present formally the business case to the appropriate management groups and demonstrate the effectiveness of the technique in alleviating the historical problems of requirements definition. As is true with any project, a plan is required to organize and control the effort.

6.1   PROBLEM AWARENESS

Before one can obtain management concurrence and approval to a plan to demonstrate the effectiveness of prototyping, information systems management must clearly acknowledge the requirements definition problem and display an interest and commitment to solving it.

To introduce new technology successfully, management must have a strong motivation to alter the status quo. The introduction of any change is unsettling and resisted by many members of an organization. Change of any kind is the exception condition and orthodox prespecification is well established in most organizations. Many members of the organization will be both positionally and professionally committed to the prespecification approach. Prototyping may be viewed both as a professional threat and another empty buzzword. These people will have to be convinced that prototyping offers a new and superior approach to solving their development problems.

The best way to overcome the initial inertia and resistance of an organization is to suggest prototyping as a direct response to an immediate problem that management is concerned about. Nobody will get very excited about a suggestion on how to fix something that works. Prototyping should be presented in the context of a practical and promising solution to current and pressing problems.

Given the state of the development process in most companies, management is only too aware of the problems of obtaining good requirements. They are also probably dissatisfied with the actual payback obtained from their investment in modern prespecification techniques. It should be easy to identify high visibility projects that are exhibiting excessive churning and rework due to poor requirements definition. These pressing problems can serve as the examples against which to present the prototyping alternative. Once management's interest is aroused, sufficient resources and funding can be formally requested to underwrite a formal proposal.

6.2   PROJECT PROPOSAL

The purpose of this document is to provide a formal proposal and evaluation plan to information systems management. It provides a complete business case in support of the concept and provides a plan on how a project should be organized to pilot and formally evaluate the concept.

This proposal serves two specific functions. First, it provides a reference that can serve as the base document for management discussion of the topic. Second, it solicits management approval and commitment to a plan for evaluating prototyping. The execution of the approved plan will provide the opportunity to prove the case.

The format and structure of a project proposal is unique to each company. One would expect, however, that any proposal for prototyping would contain the following five core subjects:

The Business Case. An analysis of the current problems with requirements definition within your environment and an explanation of why prototyping should alleviate them.

The Prototype Life Cycle. A detailed explanation of the prototyping concept and how it would fit into your SDLC.

The Prototyping Center. A statement of the proposed resources required to staff and pilot a miniature prototyping center.

Project Plan. A plan outlining all the necessary steps and resources required to pilot and evaluate prototyping.

Project Economics. A statement of expenses associated with executing the project plan.

Each of these subjects will now be discussed in more depth.

6.2.1 The Business Case

The purpose of this part of the proposal is to explain why prototyping is a conceptually appealing answer to the definition problems encountered by the organization. Chapter 2 provides the general arguments that can comprise the business case.

The arguments that are presented in Chapter 2 are too general in content and should be anchored in actual organization case problems. By showing concrete examples where the inadequacies of the current definition approach resulted in expensive rework or failure, the case is obviously made much stronger.

6.2.2   The Prototype Life Cycle

The purpose of this part of the proposal is to explain the mechanics of the prototyping approach to requirements definition. Management is certainly aware and procedurally comfortable with the in-place methodology. It should be explained how prototyping is actually done. Chapter 3 provides a reference from which to tailor a custom set of procedures.

Many people, especially those who are historically wedded to prespecification techniques, automatically equate prototyping with undisciplined and chaotic rapid development. The presentation of the prototype life cycle as a structured and phased definition approach that addresses all the definition issues is important to demonstrate the practicality, completeness, and maturity of the concept.

6.2.3   The Prototyping Center

Staff, hardware, software, and project room resources are required to perform the prototyping. The general requirements for a prototyping center are outlined in Chapter 4. Though a complete facility with large-screen projector and local batch printing may not be realistic to request for a minicenter for piloting, certainly most of the resources, in proportion, need to be provided.

This part of the proposal should include a recommendation on a prototyping software architecture. This requires that the proposal team perform a vendor search to identify a software architecture that can support rapid prototyping.

Many of the critical requirements for such software, dictionary driven, supports component engineering, provides a prototyper workbench, provides automated documentation, and so on, are provided in Chapter 4. These general requirements need to be customized to your particular environment. For example, custom requirements may address the following points:

The software must run on an established hardware configuration or a new hardware architecture is acceptable.

Vendor	Software Architecture
Cullinet	IDMS
Applied Data Research	Datacomm/IDEAL
Adabas	Natural
Burroughs	LINC

Figure 6.1 Potential vendors for prototyping software. Many vendors offer products that can be considered for use as prototyping software.

The software must be able to have good potential as a production architecture as well as a prototyping architecture.

If the software is on a new hardware architecture, it must have various communication and interface facilities to the installation's primary production architecture.

The sum of these requirements becomes the definition baseline for evaluating software.

Unfortunately, it is not possible to look up "prototyping software" in the various software reference books. Rather, those vendors that advertise their software as integrated and high-productivity oriented will have to be researched to establish an initial review list. Care must be taken in evaluating alleged "integrated" products. Many products that allege to be integrated software systems upon close examination are only integrated on the bill you pay. Figure 6.1 lists some vendors that could be considered research starting points. None of them will offer a perfect match to the requirements but depending on your particular approach, they will at minimum provide a good educational experience about what is possible and at best provide a viable candidate architecture. Of these, the IDMS architecture by Cullinet is clearly the most comprehensive product for prototyping complicated applications.

In addition to the requirements baseline for the prototyping software, the evaluation procedure should include one or two benchmark problems. Benchmarks are historically used as a means of evaluating competitive products from a performance perspective. In this case, it is desirable to use benchmark problems to compare and evaluate the efficiency of the software in enabling rapid prototyping. We would like to test the product's ability to in fact build and change representative prototype problems.

Requirement
1. Integrated Data Dictionary
2. Integrated Software Architecture
3. Menu/Command Driven Software
4. Component Engineering
5. Cut And Paste
6. Specification By Declaration
7. Prototyper Workbench
8. End User Facilities
9. Automated Documentation
10. Terminal Support
11. Data Base Manager
12. Support Virtuoso Performers
13. Security Features
14. Screen Generator
15. Automated Tunnel Editing
16. Structured Programming
17. Late Binding
18. Debugging Aids
19. Training/Documentation
20. Versioning Support
21. Logical Records
22. Dynamic File Declaration
23. Production Center Considerations
24. Other

Figure 6.2 Benchmark evaluation criteria. The benchmark solutions should be compared with the software requirements to determine suitability.

Good benchmark problems are important because they separate fact from fantasy. All vendors claim that their software is extremely flexible and enables dramatic increases in productivity. A benchmark provides a clear way to evaluate the claims. A checklist similar to Figure 6.2 can be used to evaluate the candidate software against the requirements.

Since the software being considered may require considerable training and skill building before your own staff could build the benchmark models themselves, the vendor may have to do the actual construction. This reduces to some degree your ability to evaluate the product. Consequently, an important part of the benchmark process is performing iteration on the benchmark where your staff acts as disgruntled user. The ability of the vendor to make a wide variety of changes to the problem in a reasonable time frame is highly indicative of the malleability of the software. A group of heroic vendor system engineers may in fact be able to build a model rapidly. Maintaining its integrity under rapid iteration will only be possible if the previously itemized software requirements are met. The ability to revise the benchmark successfully under simulated conditions is the acid test for a candidate software architecture.

A Benchmark Example

An example of a representative benchmark problem is the creation of a generic HELP system. A HELP function is required by almost every on-line application. In practice, the absence of a run-time HELP severely compromises the user friendliness of an application. The construction of a HELP system is an ideal benchmark problem from two perspectives:

It is a highly structured problem with a predominance of "tunnel editing" and record management function.

It is a highly familiar and practical problem. It is easy for you to act as discriminating user and easy for the vendor to understand without extensive analysis.

A HELP system should consist of three primary subsystems:

A record maintenance subsystem, on-line, for the maintenance of all system records.

A run-time subsystem, on-line, that permits the accessing of the HELP data base as required during host system execution.

A data base reporting subsystem, batch, that permits the periodic reporting from the data base of the record content and relationships in various formats.

The goal of HELP system is to provide a generic solution for these functions and be "pluggable" by a simple interface to any host prototype. HELP would be a reusable subsystem, the most beneficial type of component engineering. It would dramatically improve prototyper productivity and at the same time deliver tremendous function to the user.

Figure 6.3 is a possible normalized data base structure to support a generic HELP system. A HELP system would deliver the following type of functionality:

1. At run time a request to perform HELP record maintenance would transfer control from the prototype host to HELP. HELP would manage the record maintenance function. At the completion of the maintenance activity, control would return to the host at the point of invocation.

2. At run time a request for run-time help would transfer control from the host to the HELP system. The control transfer would deliver the user

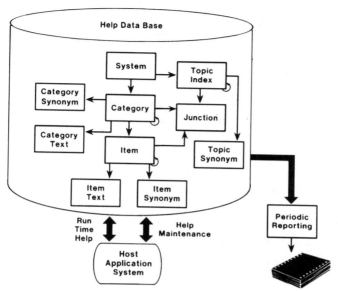

Figure 6.3 Sample benchmark problem. A generic HELP subsystem can serve as a good benchmark problem.

to a general HELP menu or to a specific HELP item if a specific HELP category item was passed. Once in HELP, the user could display, browse, and search through the HELP records until all the HELP wanted on any categories or items is obtained. At the completion of run-time HELP, return would pass to the user at the point of invocation.

The problem, though conceptually simple, requires a great deal of record manipulation and maintenance of inter-record relationships. A "SYSTEM" may own many HELP "categories." A category could be "commands" or "messages." A category could

☐ have an unlimited number of synonyms
☐ have unlimited text explaining the category
☐ be decomposed into subcategories. Each subcategory could also be further decomposed
☐ own detail help "items"

An item represents a specific subject for which help is available. Examples would be "message: 246341" or "command: display customer" where message and command are categories and 246341 and "display customer" are items. An item could

- [ ] have an unlimited number of synonyms
- [ ] have an unlimited amount of text explaining the item
- [ ] be related to other items. An item is able to point to another item and say "for related information see"

A SYSTEM may also own a "topical index." A topical index provides logical groupings of categories and items. A topical index may

- [ ] have an unlimited number of synonyms
- [ ] be decomposable into a subindex
- [ ] point to many categories and items

The maintenance subsystem would require at minimum functions such as ADD, CHANGE, DELETE, DISPLAY, and BROWSE. The problems included with equating synonyms to their owners, maintaining the cyclic bill of material structures for category, item, and topical index and completely rippling the effects of a DELETE function demonstrate the functionality and power of the data base.

The run-time HELP subsystem would require extensive SEARCH, LOCATE, DISPLAY, BROWSE, and SHOW functionality. Having entered HELP at some point, the system should aid the user in navigating all the possible relationships until the needed data is found. For example, suppose a user is looking at a specific topic index. They would now like to look at all the related subindexes. Having found the subindex of interest, they would like to see all the related categories. Having picked a category, they would now like to view all the subcategories. Having chosen a subcategory they would now like to see all the items covered by that subcategory. For a chosen item they would like to browse all the explanatory text. They would now like to look at all the items related to this item. And so on and so on. A run-time HELP needs to provide navigational aid to help the user find the ultimately needed information.

In addition to the record manipulation problems, numerous data elements of all types requiring various editing could be distributed across the records further enriching the functionality required of the HELP system.

In summary, a HELP system can provide a good benchmark. Though conceptually straightforward, it requires a great deal of functionality to be of value. Each benchmark can be analyzed to assess the delivered products compatibility with the software requirements.

Measuring the Benchmark

In the event that multiple architectures look promising it would be advantageous to supplement a purely qualitative assessment by attempting to quantify the software. It would be superior to have a metric, even if only a gross measure, which would indicate the congruity of the software with the requirements.

Considering the constant product of a benchmark situation, lines of code may serve as a gross measure of software compliance to requirements. In the benchmark situation, function and presentation are constants. Two of the primary requirements for prototyping software are reusability by component engineering and specification by declaration. Both these tactics have one clear goal—the elimination/minimization of coding. Given a constant deliverable by competing vendors, a gross difference in lines of code should be explainable to a large extent by how successfully these two tactics are implemented within each proprietary architecture.

Not all lines of code should be counted. Figure 6.4 summarizes the typical types of commands that appear in a procedural language. The general classifications are as follows:

Data Manipulative.   These are commands to manage the editing, assignment arithmetic, and comparison of data.

Record Manipulative.   These are commands that control the input/output manipulation of the data base by the program.

Screen Manipulative.   These are commands that permit run-time control of the attributes and variables associated with a screen.

Interprogram Control.   These are commands that control the transfer and return of run time control between programs.

Utility.   These are commands that facilitate debugging, logging, recovery, and so on.

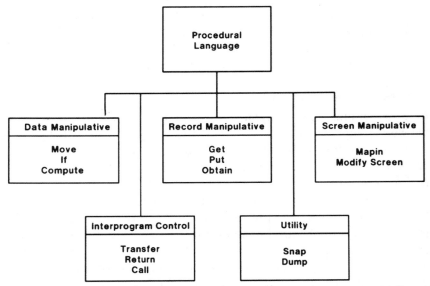

Figure 6.4 Procedural language command structure. An analysis of the delivered source code can aid in determining the degree of component engineering and specification by declaration of the candidate software architecture.

One would expect that an architecture which has high reusability or has off-loaded coding entirely by declarative specification would have an absolutely minimum amount of data manipulative commands. Consequently by counting and comparing the data manipulative commands included in each benchmark, one should get an indication, even if only gross, of the relative componentry and declarative specification delivered by each system.

If you have trouble intuitively accepting this, consider the types of logic structures that lend themselves to off-loading. To be a good candidate for declarative specification, the function has to be both highly repeatable and well defined. These characteristics are primarily true for editing and formatting logic.

Consider then an application that has to input/output to screens multiple numeric fields. The following global specifications are given for these fields:

The fields vary in size from 0.9999 to 999,999.

The operator should be able to enter the first character of the field anywhere within the field's limit.

Leading zeroes should be suppressed on output.

Dollar signs and commas may or may not be inputted but should always be present on output.

Different fields have different range checks.

Some fields are required on input, others are optional.

An error message with the name of the field should be outputted if the field is in error. The field should be highlighted.

A specific field would be a customized combination of these requirements.

If a prototype had 30 fields that had to be edited to those specifications, procedural solutions where all the logic had to be coded would result in massive amounts of code relative to solutions where the editing could be declared. A declarative solution might look as follows:

- ☐ FIELD: salary, PICTURE: $$$,$$$.99, EDIT TABLE: checksal, REQUIRED: yes, ERROR MESSAGE: 90026, ERROR ATTRIBUTE: high intensity
- ☐ FIELD: bonus-pct, PICTURE: ZZ.999, EDIT TABLE: bonus, REQUIRED: no, ERROR MESSAGE: 90082, ERROR ATTRIBUTE: high intensity

This type of solution for 30 fields requires no procedural code for editing. An architecture that did not support such specification would probably require 1000 or more lines of data manipulative code. A comparison of the data manipulative lines of code can consequently serve as a reasonable surrogate measure or indicator of the relative componentry and declarative capabilities of the architecture.

6.2.4 Project Plan

The project plan lays out the steps and resources necessary to:

- ☐ create a miniature prototyping center
- ☐ perform pilot projects
- ☐ evaluate the results of the pilots

It is the approval of this plan that is being requested by the proposal. Execution of the plan will demonstrate the benefits of prototyping and provide the justification for establishing a formal and on-going prototyping function.

6.2.5   Project Economics

The cost of executing the evaluation plan should be estimated. At minimum, the following cost items would be expected.

Labor.   This includes the loaded cost of prototypers for the duration of the project.

Education.   This includes the cost of training each prototyper in the selected software architecture.

Hardware/Communications.   This includes the cost of any new hardware/communication equipment required for the pilot center.

Machine Resources.   This includes anticipated computer expenses.

In addition to these one time piloting expenses, management should be made aware of any long term but not obvious expenses associated with prototyping. Examples of such expenses might be the purchase and yearly maintenance cost of the software, the creation of a pool of personal computers, or acquisition of a large screen projector system.

6.2.6   Plan Approval

There is every reason to anticipate that a well presented project proposal will be enthusiastically accepted by management. The current process, prespecification, has at best mixed results. The intuitive appeal of prototyping when supported by a well prepared proposal should be sufficient to make even the most skeptical management not want to risk missing such a potentially rewarding opportunity.

6.3 EXECUTING THE PLAN

A prototyping project is a high visibility project. The promise of prototyping has tremendous implications in revitalizing the whole development
process. In addition, unlike its predecessor's approaches, it promises to
deliver working models. Everyone will be able to review and judge for
themselves the product. There are three steps to the plan; creation of a
miniature prototyping center, piloting, and results evaluation. The activities of each step will now be discussed.

6.3.1 Creation of a Miniature Prototyping Center

As was presented in Chapter 4, a prototyping center consists of four
primary resources: staff, software, hardware, and work environment. The
considerations for each are as follows.

Staff

A staff of two to three people with careful selection of the lead prototyper
who will serve as system architect should be sufficient. This will permit one prototype to be done at a time. The staff should be fully trained
in the software architecture so that they can function as independent vir-
tuosos.

Adequate time should be provided between the completion of the education sequence and the initial prototype to permit skill building. A number of productivity studies have labeled a deterrent to productivity called
"break through technology." Break through technology is any technology being used for the first time by a development staff. Education does
not equate to experience. The prototyping staff should suffer the experience pains of break through technology on skill building play problems,
not the real prototypes.

In addition to technical skill building, adequate interim time between
education and the first pilot is needed to create the work environment.
Only by actually building a play system, can the beginnings of a repeatable process be devised, standards suggested, and the strengths and
weaknesses of the software be assessed. In essence, the prototypers need
to throw one away before they're ready to do their first real pilot.

Software

Arrangements have to be made for installing or time sharing the selected software if it is not an already established in-house architecture. In the event that a time-sharing approach with a vendor is selected, special emphasis should be placed in the contract on required response time. Since good response time is critical to the success of the technique but it is out of your internal control, the vendor must be contractually motivated to provide satisfactory response.

Hardware

A one-to-one ratio of terminals to prototypers should be provided.

Work Atmosphere

A project room approach is absolutely mandatory. The high productivity environment that was described in Chapter 4 is a key component of the entire process. Though perhaps difficult to justify for a pilot project, the availability of the large screen projector would be highly advantageous. The large screen projector makes a tremendous difference to the iteration steps with the users.

6.3.2 Piloting

Concurrent with the creation of the miniature prototyping center, initial projects to undergo prototyping can be selected. Depending on the amount of convincing that will be required one to three pilots should be adequate.

Care should be taken to conservatively pick the initial candidates:

The projects should be large enough to prove the point but not strain the abilities of a new staff.

The projects should be important (visible) but not so critical that failure would have any major negative effect on the company.

The projects should clearly meet all the candidacy guidelines.

The users and project managers should be positively inclined toward the process. If necessary to minimize user concerns, a contingency plan could be provided to provide an easy exit if the user became dissatisfied.

The project should be small enough to be completed in the low side of the 2–6 week initial build period. It is highly advantageous to deliver results as quickly as possible.

For each project, success/evaluation criteria can be developed. Though quantitative measures are theoretically possible, since no one knows how to measure efficiency, yet the more difficult effectiveness attribute, qualitative criteria would be most appropriate. The best success criteria, of course, will be the enthusiastic postproject endorsements by the users, project managers, and developers.

The actual project piloting consists of executing the prototype life cycle. Each project could be completed within an 8–12 week period from needs analysis through documentation. In theory, one could (should) wait until the actual system has been built and is operational to evaluate the results fully. This is simply not practical. The target system of the piloting may take another year to build and convert. Then, waiting another year to accumulate maintenance statistics would prove academically interesting but impractical. Practicality requires that the completed prototype serve as the evaluation object.

6.3.3 Results Evaluation

The results of each pilot prototype should be documented and serve as the basis for a prototyping center recommendation. Some of the evaluation is simply factual in nature. Statistics on each prototype, number of screens, elements, programs, and so on, can be documented. Most of the evaluation is subjective. Each of the development participants, user, project manager, prototypers, and developers, should express their evaluation of the process. Special attention and consideration should be given to the user's opinion. The whole purpose of any definition procedure is to meet the user's needs. The user's satisfaction with the process is much more important than that of any other participant.

The evaluation reports should be presented to management with a go/no-go recommendation. Assuming success, another project will be required to integrate prototyping into the SDLC as a standard and fully funded development function.

6.4 MARKETING

Introducing new technology requires salesmanship and marketing skills. Concurrent with the creation of the project proposal and the subsequent piloting, prototyping has to be aggressively sold to the organization. The technology has to be made a welcome addition to the organizations culture. We want everyone and anyone to be informed and talking about it. This can be accomplished in many ways:

Presentations.   These are formal presentations explaining the project proposal.

Newsletters.   These are communications to the user and development communities explaining the concepts.

Testimonials.   These are guest speakers from other companies that have implemented prototyping.

External Consultants.   These are the opinions and experience of a respected third party.

The expertise of many data processing leaders can also be used to sell prototyping. Respected and highly recognized industry leaders such as James Martin and Daniel McCracken strongly endorse prototyping. The distribution of their writings on the subject add considerable weight and credibility to the marketing effort.

Many of the largest software vendors are also actively endorsing prototyping. Respected companies such as Cullinet, Applied Data Research, IBM, Hewlett Packard, and Cincom Systems all recommend prototyping in their current marketing literature.

In summary, there are numerous mediums for marketing the prototyping concept and considerable industry support to reference to support the sales effort. A consensus will be required to make formal prototyping a standard offering and preferred technique. By making as many people as possible informed and comfortable with the technique, the greater support will make the final management decision obtained more easily.

6.5   SUMMARY

There is, unfortunately, a great deal of inertia in data processing organizations. Change does not come easily but positive change is critical to
accrue massive productivity improvements. The sad fact is that as an
industry we routinely deliver prototypes as real systems. They are certainly not called prototypes but they exhibit all the imperfections of initial
models. To change this expensive and wasteful process, prototyping must
be introduced into the organization. This requires a well prepared business case and plan. Many good ideas never reach fruition in the data
processing world because of inept introduction and lack of commitment.
By carefully planning its introduction, staging visible and successful
pilots, and constantly marketing the concept, the inertia and skepticism
can be overcome.

SEVEN

ISSUES
AND CONCERNS

The introduction of prototyping as an alternative definition strategy will raise a lot of questions. Segments of the user, project management, auditing, and developer communities will all have many legitimate questions about the process and its impact on the SDLC. The best way to alleviate the concerns is by being prepared to answer them.

The purpose of this chapter is to provide responses to common concerns that are raised about prototyping. The following format will be used:

Concern. This will be a brief topical statement of the issue.

Discussion. This will be a narrative explanation of the perceived problem.

Response. This will be a narrative explanation that answers the stated problem.

Prototyping is neither a perfect approach nor is it always appropriate. Concerns to which a response is unsatisfactory might indicate situations where prototyping is inappropriate and prespecification would be a better primary strategy.

1 *Concern*

Prototyping glosses over the real business problems.

Discussion

Prototyping jumps much too quickly to building a base solution. It is likely that only superficial and obvious needs will be addressed. Most of the complicated problems will be glossed over. Only a detailed and thorough analysis of the business problem prior to any physical solution can assure that all the problems will be addressed. Prototyping, in practice, will result in fast but expensive false starts.

Response

The real problem is not whether you may initially miss some needs but how you will strategically discover all the ultimate needs while minimizing project risk. Descriptive and graphic prespecification tools are fine but are extremely limited in portraying the actual dynamics of a potential application. Prototyping provides a means to, by intent, first approximate the real needs and then through successive refinement pinpoint them exactly. Even, if some needs are glossed over in the early presentations, the process is self-correcting. The contextual testing of the prototype will reveal any functional deficiencies. False starts should quickly and cost efficiently be discovered. It must be understood that this concern implies that current analysis techniques are in fact successful in predefining the business problem. This implication is at best highly argumentative. The growing industry interest in prototyping is a direct result of the failure of current techniques to adequately specify the business problem. Prespecification, not prototyping, is often guilty of glossing over requirements.

2 *Concern*

The user will want to prototype as the actual system.

Discussion

After a number of iteration cycles, the prototype will appear to the user as a finished system. If they insist on implementing it, the consequence will be an expensive life cycle product. Prototyping will introduce a whole new set of political problems into the life cycle. Users will be dissatisfied and disgruntled that they may still have to wait a considerable time until they actually get the system.

Response

A prototype is exactly what its name implies: a model. There is no magic. Prototyping provides an illusion. It gives the impression of a finished system but it is missing many of the attributes of a production application. It is incumbent on the prototypers during candidacy selection to clarify the limitations of a prototype to the users before starting the project.

Nobody insists that descriptive/graphic definition documents be put directly into production. Similarly, a prototype must be first and foremost viewed as a definition document. Cheating the process will not work. Reality will detect every shortcut taken in attempting to implement a model directly.

Though not directly implementable, the prototype can still help the user during the interim period. If appropriate, it can expedite the design of the actual system by serving as the basis. The prototype can also serve as a training system for the users.

The political problems of dealing with a user who insists on implementing her or his prototype system is really a rather pleasant change. Most data processing managers have to deal with users screaming about how dissatisfied they are with their new system rather than demanding implementation because of satisfaction.

3   Concern

Prototyping is a departure from established software engineering practices.

Discussion

The body of knowledge developed within the profession over the last few years has focused on rigorous approaches as the best response to development problems. Many software engineering principles and techniques are implemented within our organization. Endorsement of a prototyping approach will jeopardize the attempt to make software development an engineering-like process.

Response

Quite the opposite is true. The building of models is the central step of most engineering life cycles. Most engineering disciplines gave up a long time ago attempting to do it perfectly without building some laboratory throwaways. Introducing prototyping into the development cycle will for the first time migrate the wisdom of the need for the engineering prototype to software development.

4 Concern

Prototyping prejudges a physical solution and will constrain the developers. Prototyping mixes up the what problem (logical problem) with the how problem (physical solution).

Discussion

Prototyping jumps to physical solutions without a careful logical model of the problem. Logical modeling provides a mechanism for testing various possible solutions before committing to a solution. By keeping requirements definition at a logical level, we maintain maximum implementation flexibility.

Response

For whose benefit. The idea that anyone other than the logical model creator is thrilled with logical designs is a myth. Users relate best and naturally to physical examples. When you buy a television, you look at the screen picture, you do not look at the wiring schematic.

The final physical model can be dissected at the end of the prototyping to design exactly many of the system components that logical modeling normally develops. For example, the necessary logical views of the data base can be predicted exactly by analyzing each screen. The secondary keys that are required can be exactly predicted by analyzing all accesses. The logical guessing can be replaced by physical confirmation.

The user's view of the application is still an external view. There is nothing to prevent, as appropriate, the developers from revising data base structures, using a distributed data processing solution, or selecting different access strategies. When physical realities will require compromises, it will be clear what the before and after are. The user's firm insistence on the demonstrated user/machine interface is what the developer has always needed and wanted: a firm and nonvolatile definition.

5 *Concern*

The availability of qualified prototypers is extremely limited.

Discussion

Even if conceptually and operationally successful, there would be a shortage of individuals able to be application architects. The methodology does not allow for an adequate division of labor. The ability to offer prototyping will be severely constrained by the need for highly skilled prototypers.

Response

Analysis has to be done. The availability of a competent analyst for prespecification is as much a constraint as the availability of competent architects for prototyping. When performance and operational constraints can be ignored, the building process directly equates to the business problem. It is not unreasonable for people who can conceive logical visions of an application to do one more step and make that vision operative.

The skills of the prototypers only have to be proportional to the problem. Working for an extremely large company, I tend to view the data processing problem space as filled with applications of the two sizes: large and larger. My view of the prototyping environment naturally requires highly skilled architects to address such problems. In practice, the problems that are presented for prototyping will be distributed in size and difficulty. The staff, likewise, needs at any time only to be skilled in synchronization with the normal workload.

Again, reality is the critical test. If descriptive models are inadequate, there is no choice. You can continue a farce or take the necessary actions to staff a prototyping unit. Nobody ever promised that building applications will be easy and able to be done by everyone. Specialists are needed in all disciplines to do a quality and competent job. Prototyping of large applications requires specialists to do the skilled work just like any other profession.

6 *Concern*

The absence of logical modeling will stifle the creation of innovative solutions. Prototypes will tend to translate the current process with all its inefficiencies into the new system.

Discussion

One of the major benefits of structured analysis techniques is that the business process can be abstractly modeled. The abstraction of a logical model permits the opportunity to suggest and test novel ways of doing the business.

Consider Figure 7.1. First a physical model of the current environment is created. It is then abstracted to a logical equivalent. The logical model of the current environment can then be massaged by the feasibility study, analyst creativity, user insight, or any other factor to derive multiple candidate models of the new logical environment. The best logical model based on consensus agreement can become the selected solution.

The creation of the logical models of the new environment provide an opportunity to innovate. The flow of the whole system can be revised to yield a much more efficient set of procedures. The jump to solution nature of prototyping precludes this opportunity for such innovation.

Response

This is probably the best argument in favor of prespecification techniques. The concept of abstracting the problem, massaging it in multiple ways, and then choosing an optimum solution is a powerful concept.

There are a few points to be considered, however, before conceding this concern in its entirety:

1. Nothing precludes performing logical modeling during the needs analysis of prototyping. If the current process is in need of a clear overhaul, the best of "rigor" certainly can be used during prototyping to help analyze and develop a first solution.

2. Creative analysis of the kind required to alter a current environment radically requires an exceptional analyst. Given the labor intensiveness of structured analysis and the real world time constraints on an analyst, it is quite a job to get out a new logical model that has hope of working yet alone multiple proposals.

3. Prototyping creates a very exciting environment to stimulate creativity and innovation. Users can suggest in brainstorming sessions novel ways they might like to do their job. The best ideas can be demonstrated quickly. At the extreme, multiple models of a proposed system can be created, demonstrated, compared, and evaluated. The primary catalyst to innovation is people, not methodology.

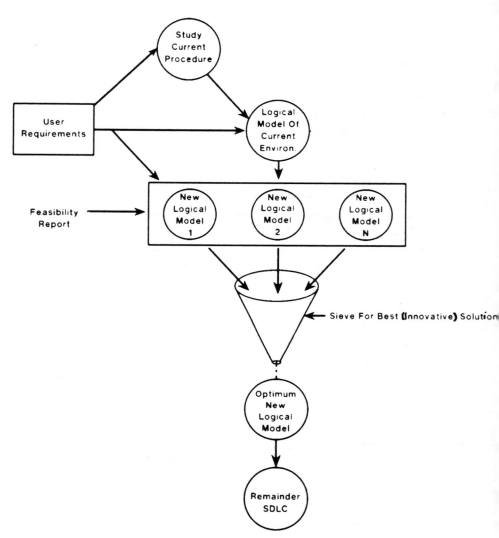

Figure 7.1 Logical system modeling. Provides a potentially rich opportunity for innovative solutions.

4. Most systems are rather "vanilla." They neither warrant nor need particularly novel solutions. In fact, given the user's historical dissatisfaction with information systems, they would find any workable solution a blessing. More often than not, the elaborate and elegant solutions have to be applied to the data base design and performance issues that the users don't appreciate at all.

In summary, the ability to model multiple visions of a new environment abstractly is an undisputable benefit. Though potentially a powerful tool for generating novel solutions, practical time, skill, and problem constraints limit the delivery of such innovative solutions. Prototyping offers ample opportunity for innovation by permitting ideas to be put to operational testing.

7 *Concern*

Prototyping appears to be much more formal and difficult than anticipated. The impression given from prior reading was that it was simple, fast, and easy.

Discussion

Prototyping as you present it is a substantial undertaking. Skilled prototypers, special software, and a customized work environment are all required. It was my impression that the whole process would be much simpler and quicker. Is such an elaborate undertaking really required?

Response

If we were to rephrase this concern, it would read: Why doesn't prototyping deliver magical solutions? The answer of course is that the sum total of magic involved in building applications is zero.

The impression that somehow prototyping would be an extremely simple effort is due in part to the three level consultant problem. The first level consultant, an extremely high-level guru, suggests prototyping at the conceptual level, answers a few questions, and leaves. The second level consultant, who almost has his or her feet on the ground, will confirm the wisdom of the initial guru and help in preparing the project proposal and selection of software. His or her work is now done. Now comes the third level consultant. This individual will actually be able to help do the prototyping.

Most of the interviews, articles, books, and seminars are given by the level one and level two consultants. They naturally tend to extoll its virtues and minimize its challenge. The third level consultant is not available to talk much about prototyping; he or she is too busy doing it. Though creating the philosophical framework and support for prototyping is certainly important, it is one thing to talk about it and quite another to have to do it.

Consequently, the available literature tends to focus on the concept and justification of prototyping as opposed to the operational procedures required to carry it out. Unless you are going to be doing only trivial problems, most of the requirements that have been suggested are needed. If the prototypers are not highly skilled, how will they deliver a model rapidly? If the software is not dictionary driven and supports component engineering, automated documentation, specification by declaration, and so on, how will the prototypers have sufficient leverage to do the work? If the prototype life cycle does not include provisions for specifying rigorous components, when will they be specified?

8   Concern

The handoff from the prototypers to the actual developers will be difficult.

Discussion

The documentation language for prespecification techniques is in English. The developers can read it without special training. The prototype will be delivered to the developer in the language of the prototyping software. Additional expense and training will be required to facilitate the transfer between phases.

Response

This is true. A working prototype is a completely unambiguous specification. The prototyping software acts as a technical specification language. The developers must have reading level literacy in the language. If the prototyping architecture is in fact the final target architecture, this should be no problem. In the event that the target architecture is different, some literacy training to facilitate translation will be required.

The participation of a lead developer as part of the prototyping team should help in minimizing translation problems. Of course, the prototypers and the animated model are also available for help.

9 *Concern*

The prototype is unconstrained in the functionality it will demonstrate to the user. The user will be dissatisfied when the real system cannot deliver the entire model.

Discussion

The advantage that the prototypers have in ignoring the physical attributes of the application is a luxury the real system will not enjoy. Function demonstrated to the users on a small data base may very well prove impractical or too costly in the target application. Users will be highly dissatisfied that the model cannot be delivered or that the actual system will be prohibitively expensive.

Response

The prototype is a requirements document. Either the demonstrated requirement is necessary to the business operation or it is not. A needed requirement can be met either manually or mechanically. If the target architecture cannot deliver a needed requirement, then the problem is the target architecture. If the problem is cost, what you are really saying is that it is cheaper to do the required function manually. If the cost/benefit ratio does not support mechanization, then the function should be removed from the prototype but exactly and at what cost it will be done manually should be explained.

The rigorous definition step of the prototype life cycle should catch any constraints that operationally will invalidate the model. One would expect that given the inspection opportunity which a prototype offers to all members of the development community, an unimplementable prototype should be a rare occurrence.

10 *Concern*

The iteration process could go on indefinitely. How do you prevent iteration from thrashing?

Discussion

Prototyping is a license to change one's mind. Looping could occur indefinitely between demonstrations and revisions. The process could become extremely drawn out. After weeks of demonstrations, we might not be closer to an acceptable model than when we started.

Response

This is exactly so. Not only won't you be close to a viable solution but also you and everyone else will know it. One of the primary benefits of prototyping is that it serves as a high visibility safety valve. A project that is going nowhere in prototyping is a project which requires close management attention. If agreement and consensus cannot be reached now, better to scrap the project than pour huge sums of money into an undefinable problem.

11 Concern

Prototyping conflicts with data resource administration and the creation of shared data bases.

Discussion

Data needs to be treated as an independent resource. The definition, cataloging, structuring, and security of data needs to be done independently of the functional applications that use it. The result of such an approach will be the creation of a shared data base resource where multiple application access the same data (see Figure 7.2). Prototyping, since it concentrates only on a specific application, will tend to take a tunnel view of the world, and conflicts with the greater issue of proper data and data base administration.

Response

There is only a possible conflict if a complete development life cycle is not completed and the prototype is rushed into production as an isolated application. During analysis, the current base of data and data bases in which the application must be integrated should certainly be considered. If many of the data elements, records, and relationships are already

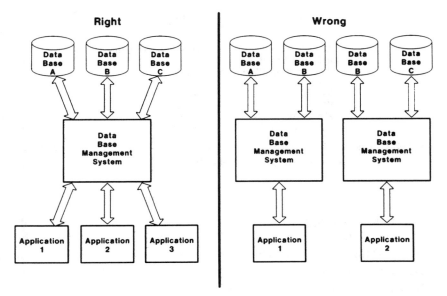

Figure 7.2 Data base administration. Proper data and data base administration will result in shared rather than application only data bases.

defined, more of the work is done before we ever start. One of the primary reasons for maintaining the SDLC after completing the prototype is to consider properly the integration of the application into the overall data and data base plan.

In fact, treatment of established prototype data base as a shared and reusable resource yields the same benefits to the prototypers as the production data base libraries do for the actual developers. Elements, records, and relationships are already defined and reusable. Only the new and exceptional aspects of the application need to be developed. Proper data administration should not be contrary to effective prototyping and in practice should aid and expedite the process.

12 Concern

There are many applications that are not structured and transaction oriented. How do I increase my productivity in addressing those problems?

Vendor	Product
IBM	ADRS
CULLINET	On-Line English
Information Builders	Focus
Artificial Intelligence	Intellect
UNIVAC	Mapper
NCSS	Nomad
IBM	QBE
MATHEMATICA	Ramis
VARIOUS	APL/BASIC

Figure 7.3 End-user software. Unstructured problems are best served by end users themselves, developing and evolving the application.

Discussion

Transaction based applications such as inventory control, billing, order entry, and contract administration, make up only part of the applications portfolio. Much of the growth in new applications is in the query and analysis area. If prototyping is inappropriate for the types of problems, how should they be addressed?

Response

As was stated before, most conventional applications can be viewed as "the business" or "about the business." Prototyping is required for "the business" systems because of their size, complexity, impact on the business, high uncertainty, and risk of failure.

Much of the growth in new applications is in the about the business group. Systems that are about the business can be labeled under the generic titles of information retrieval, data base inquiry, graphics generation, report generation, decision support, and personal data bases. What is true about these applications is the volatility and unstability of the requirements. The systems are inherently dynamic prototypes. They need to be highly flexible and responsive to constantly changing demands.

It is not so much that a third party prototyper could not build an initial model of such an application as that, these types of applications are better built directly by the end users. Since the only predictable and stable component of such an application is the data base, after helping define the appropriate views of the data base resource, it makes the most sense to teach the user how to build her or his own solutions. Given the improving state of end-user computing facilities (Figure 7.3) end users can become highly self-sufficient in developing these types of applications. Prototyping will never end for these types of applications due to their inherent nature. The best strategy to address them is to support the end user as developer.

Suggestion

It is very effective to include in formal presentations a slide that itemizes some of these common concerns. By presenting the issues up front, not only do you generate considerable discussion but also a very positive impression is made on the audience with regard to your preparation and consideration of the method.

All these questions are reasonable and represent careful consideration of a new methodology by concerned individuals. Confidence in prototyping is exhibited by directly addressing all issues when they are raised. This includes acknowledging situations when it will be inappropriate.

EIGHT

EPILOGUE

Prototyping has always been desirable but only recently possible. The traditional analysis methodologies have not adequately come to terms with the three overriding and persistent problems of requirements definition:

Users have extensive difficulty in prespecifying final and ultimate requirements.

Descriptive/graphic modeling techniques are inadequate to portray the dynamics of an application or to insure consistent interpretation by all project participants.

Poor communication is an inherent and debilitating problem between project members.

In spite of all the well intentioned effort to systemize and discipline the definition process, it is still neither uncommon nor suprising to see projects go through extensive and costly rewrite to reconcile that which was delivered with what in reality is wanted and needed by the user. The current process often does not work. Processes that do not work need to be corrected or replaced by processes that do.

The correction to the current problem of requirements definition is the evolutionary development of requirements by the building and refinement of models. A major commitment of expensive staff and resources should not be made for new applications until an operational system model has been experienced and concurred with by all system effected users.

Prototyping is now possible because both the philosophical concepts and operational tactics needed to permit rapid construction and revision exist. Philosophically, sufficient experience has accumulated that most conventional business applications can be associated with a few simple models. The system structure, function, editing, and reporting have all been done before. Only the special nuances and subtleties of this particular application need to be discovered.

Tactically, methods and tools exist that permit the execution of a compressed system life cycle both quickly and efficiently. System modeling would not be feasible if the effort and cost would equate to building the actual system. The advent of integrated software architectures supplemented by high productivity work environments have made this a reality.

The convergence of philosophy and tactics has permitted the definition process to become animated. Users can first examine other but similar applications (systems-by-example) to develop product awareness and sophistication. Iteration can permit incremental learning and experience. All of which takes place within the contextual familiar setting of terminals and reports. The user will be a consumer who sees the purchase before acquisition.

Conclusion

Consideration of prototyping as a preferred definition strategy will result in six primary conclusions about its nature:

Prototyping is highly desirable from the user's perspective.

Prototyping is highly desirable from the developer's perspective.

Prototyping is applicable to large applications.

Prototyping is feasible.

The prototyper equates to an architect.

A prototyper center philosophy provides the work atmosphere required to enable the process.

Prototyping is a high productivity strategy for solving the real world problems of defining business system requirements. Its success is anchored in a simple yet elegant formula: start small, gain acceptance, evolve. The ability to deliver animated prototypes is a watershed event. The requirements definition problem is finally solvable.

INDEX

DATE DUE

JUL 1 1 1998			